Demystifying Climate Change

Demystifying Climate Change

An Energy Story on Science, History, Threats, and Opportunities

Jeffrey Loehr
Margaret Loehr

ROWMAN & LITTLEFIELD
Lanham • Boulder • New York • London

Published by Rowman & Littlefield
A wholly owned subsidiary of The Rowman & Littlefield Publishing Group, Inc.
4501 Forbes Boulevard, Suite 200, Lanham, Maryland 20706
www.rowman.com

Unit A, Whitacre Mews, 26-34 Stannary Street, London SE11 4AB

British Library Cataloguing in Publication Information Available

Library of Congress Cataloging-in-Publication Data

Names: Loehr, Jeffrey, 1973– author. | Loehr, Margaret, 1945– author.
Title: Demystifying climate change : an energy story on science, history, threats, and opportunities / Jeffrey Loehr, Margaret Loehr.
Description: Lanham : Rowman & Littlefield, [2017] | Includes bibliographical references.
Identifiers: LCCN 2017024656 (print) | LCCN 2017027538 (ebook) | ISBN 9781475835830 (electronic) | ISBN 9781475835816 (cloth : alk. paper) | ISBN 9781475835823 (pbk. : alk. paper)
Subjects: LCSH: Climatic changes. | Power resources—Environmental aspects. | Renewable energy sources. | Climate change mitigation.
Classification: LCC QC903 (ebook) | LCC QC903 .L74 2017 (print) | DDC 363.738/74—dc23 LC record available at https://lccn.loc.gov/2017024656

∞ ™ The paper used in this publication meets the minimum requirements of American National Standard for Information Sciences Permanence of Paper for Printed Library Materials, ANSI/NISO Z39.48-1992.

Printed in the United States of America

For Kenneth and Frida
and all children and grandchildren

Contents

Preface

Driving north from the Los Angeles basin requires traversing the San Gabriel Mountains on a winding highway called the Grapevine. This road descends into the southern end of the California Central Valley—one of the most productive agricultural regions in the world. It is a place of intense sunshine and summer midday temperatures often in triple digits.

The parched landscape just beyond the mountains consists of miles of dry grass, an occasional weary-looking steer, and dozens of rusted, tired oil derricks relentlessly bobbing up and down in the heat like so many drinking birds, extracting crude oil from diminishing stores deep underground.

The irony of this scene is palpable.

When oil seemed inexhaustible and the threats posed to the environment by its overuse were largely unknown, these derricks made a lot of sense; this was how we fueled modern civilization. We have all benefited from this incredible source of energy and the abundant lifestyle it has brought.

Today, however, the continued pumping of the last bits of ancient fuels out of the ground, in the midst of plentiful quantities of free, clean, nonpolluting energy, is the poster child for shortsightedness and irresponsibility. Not only are our supplies of oil running low and requiring increasing environmental damage to extract, but the use of oil and other fossil fuels is, without question, causing warming to our planet that carries threats to all future generations.

Over the last two hundred years humans have used incredible cleverness, scientific knowledge, and innovation to bring unheard-of comfort, convenience, and health to our species. Right now we stand at a precipice. Will we once again engage that remarkable intelligence and courage to continue and expand this quality of life? Or will we, through inaction, ignorance, and sometimes greed, lose what we have and return humanity to lives that are, as

Thomas Hobbes once said, “nasty, brutish, and short”?[1] Those of us alive today will make that choice.

Humans have an extraordinary history for rising to the occasion in the face of threats, coming together for the betterment of all, and finding the answers to serious problems. There’s no reason that we cannot do that now.

Our current global warming problems actually offer us a way forward into a healthier, cleaner, and potentially more rewarding life for ourselves and our descendants. We can have a world with clean air, clean water, and a less volatile international situation, where we could continue to live in comfort.

But first we need to understand how human activities are affecting the life-sustaining systems of the earth. Then we need to be prepared to step forward into uncharted territory with the same sense of determination that our ancestors had in the past.

This book is a story about energy, the choices we are making, and how they will affect the future. We have several goals. The first is to explain the basics of what is currently going on in the natural world. This will include an overview of the science of energy and global warming within the context of the great natural cycles and life-nurturing systems of the earth.

We will address some of the history that brought us to this place and solutions to our current crisis. What technology is available now and is foreseen for the future that can help us move away from fossil fuels? What are some economic solutions available to us? How can we preserve our privileged way of living without condemning future generations to despair?

Finally, what kind of a future can we expect for ourselves and our descendants if we take responsibility for our climate change/fossil fuel problems? If we, as a global community, begin to address the problem soon, there will be no need to consider life in a cave or take up asceticism. There is every reason to believe that we are on the cusp of a more abundant and safer world.

We begin our story by taking a look at this stuff we call energy. What is it? Where does it come from? Where does it go? And how does it affect every aspect of our lives?

Part I

The Nature of Energy

Chapter One

The Protagonist of the Story

Energy

Energy and persistence conquer all things.

—Benjamin Franklin

Sir Isaac Newton, one of the most accomplished scientists of all time, never mentioned energy. He explained motion and forces, he could predict the behavior of celestial bodies, he was spot on in his explanation of colors of light, and he invented calculus. But the idea of energy whizzed right by him. In fact, it was not a scientific concept for decades after his death.

Some 350 years later, an obscure employee in a Bern, Switzerland, patent office named Albert Einstein stunned the world when he demonstrated, with his famous equation $E = mc^2$, that energy is, quite literally, everything.

In defense of Newton, energy was not much of a problem in his time. Toss a few logs or a bit of coal on the fire, or feed the "engine" of your donkey cart a bit of hay, and you were good to go. Plenty of the stuff around. But today, the vast quantities of logs and coal used, as well as the fact that our modes of transportation no longer run on hay, can't be ignored.

Today, energy affects every aspect of modern life. Our understanding of it and our willingness to act on what we know holds the key to our future.

WHAT IS IT?

Let's start at the beginning. For those who took copious notes in school and go over them on a regular basis, this may be repetitious. If you want to move on to a discussion of things more global, check out the "takeaway points" and the end of each chapter and skip this section.

For the rest of us, however, a review of energy basics is probably a good way to get everyone on the same page.

So what, exactly, is energy, what does it do for us, and why is it so important that we are willing to send soldiers to die for it, and spend billions protecting it?

It's like love. Ask ten different people what love is and you will get ten different definitions, all of them perfectly appropriate, but none of them definitive.

The same is true for energy. Definitions abound. Walk the aisles of a grocery store and you are bombarded with the word. We frequently talk about the energy level in a room or the energy between two people. *Apple's Dictionary* takes a shot with such suggestions as "strength and vitality necessary for sustained activity, vigor, sparkle, buoyancy, pizazz, bounce, and feistiness." What's not to like? Except for our purposes none of these are useful. It's hard to imagine going to war over pizazz and feistiness.

The type of energy we are talking about is not touchy feely. There is no emotion or sense of joie de vivre attached to it. Chugging "energy drinks" or infusing oneself with caffeine is off the mark as well; that is a drug reaction—not energy.

The energy that runs our cars, computers, brains, and rocket ships is measurable, transferable, storable, and marketable. It is, to use a highly technical scientific term, a type of "stuff." Science provides an accurate, if not particularly inspiring, definition:

> Energy is the ability to do work.

"Work," likewise, has a very specific definition:

> Work is moving something.

Combining these we get:

> Energy: The ability to move something.

Therefore, energy involves:

1. Movement
2. A thing that moves

Got it?

It is worth noting up front that "work" does necessarily have a positive connotation. It can be desirable work, such as using a baseball bat to send a ball into the center-field bleachers, but it can also be undesirable, like swinging that same bat on the hood of your new car, rearranging the sheet metal. Both, in the neutral, unemotional world of science, qualify as "work."

At first blush, this may seem clear enough. Watching a 747 lumber down the runway gaining enough speed to thrust itself into the air obviously involves motion, and we can understand a lot of energy is involved, one way or another.

But what about your toaster? You know the thing is using energy; you're paying for it. But, nothing appears to be moving. A little light show, bread turned brown and crispy. Where's the motion?

Heat, whether from a toaster, your furnace, or a sunny day represents motion as much as the 747, but at the molecular/atomic level. When molecules and atoms gain energy, their motion increases—they move faster—and we sense that as heat. The more heat, the faster particles are moving. They, in turn, can give off something called "radiant heat waves," which is infrared photons—an important part of our energy story we will get to later.

The air near a heater or fireplace feels warmer than the air outside on a snowy day. The difference is how fast the air molecules are moving. Faster-moving molecules feel warmer. These faster-moving air molecules can transfer energy to your bread, speeding up the bread molecules and even causing a chemical reaction that turns the bread crisp and brown.

Likewise, your computer, phone, and assortment of energy-consuming gadgets also involve putting atomic particles, usually electrons, into motion.

So, energy has the ability to cause motion of both the very large—airplanes, cars, volcanoes—and the very small—molecules, atoms, photons.

ENERGY IN LIVING THINGS

All life forms—a flower, a sleeping cat, lichen, even the most dedicated couch potato, despite evidence to the contrary—are in constant motion. Your lungs, your heart, your capillaries, your bone marrow, every one of your cells is busy, busy, busy moving things around, taking things apart, transporting things, making new things. There is even the occasional transfer of energy in the synapses of your brain. When the doctor puts that cold stethoscope on your chest or abdomen, he or she is listening for motion. The medical term for a creature whose parts are not in motion is "dead."

Energy to maintain this cacophony of motion is more important to your body than anything else. In the event of starvation, when outside energy supplies are cut off, your body will cannibalize itself (its muscles, its organs) to maintain a source of energy. Death usually comes from organ failure, likely the heart, when full function is no longer possible because too much of its structure has been quarried for energy.

Doing things, as well as life itself, involves moving things and, therefore, requires a constant supply of energy.

WHAT IS POWER?

Often it is easier to discuss "power" instead of "energy." These terms are frequently confused, and sometimes, incorrectly, used interchangeably. They mean different things. Power is the rate of energy used, or produced. In other words, power is how much energy is used in a period of time, usually a second.

KEY TAKEAWAY POINTS

- Energy is a measurable substance that is defined as the ability to move something.
- Most things in our world, including life, are in constant motion and need a constant supply of energy.
- Power is the amount of energy per second.

DISCUSSION QUESTIONS

1. Is jumping on a trampoline doing "work"? How? What about painting a picture?
2. How many examples of energy are you using right now?
3. Can you think of something you do that does not involve motion, and therefore does not involve energy?

Chapter Two

Doing the Numbers

Measuring Energy and Power

> *Any time scientists disagree, it's because we have insufficient data. Then we can agree on what kind of data to get; we get the data; and the data solves the problem. Either I'm right, or you're right, or we're both wrong. And we move on. That kind of conflict resolution does not exist in politics or religion.*
>
> —Neil deGrasse Tyson

As previously mentioned, energy is a "stuff" that can be measured. But how? In different countries or circumstances, different units are used and that can be confusing. With a little patience, some translations, and a few numbers (not too many) we can sort it out.

QUANTIFYING ENERGY

The distance from here to there can be expressed in any number of ways: inches, centimeters, feet, meters, yards, miles, hands, light-years, and so forth. To discuss distances, any of these will do, but we need to agree on which unit we will use.

Similarly, there are different units of measurement for energy. All, however, go back to the definition of energy, which is the ability to move something; they are, therefore, based on the combination of how difficult it is to move a thing and the distance the thing is moved. It is important to understand all measurement units represent the same principle, and can be converted back and forth.

For this book we will use the modern SI (International System of Units) unit, the "joule," pronounced "jewel," named after the British brewer James

Prescott Joule, who dabbled in science while making, and no doubt drinking, beer. Note, both work and energy are measured in joules.

A joule is a very small amount of energy roughly equivalent to the work needed to move an apple from the floor to waist height. It can be translated into heat and would be about the amount needed to raise the temperature of a small drop of water one degree Fahrenheit (°F).

Other historical and specialized units for measuring energy include the Btu (British thermal unit), which is the equivalent of 1,055 joules. A unit often used in heat and food is the calorie. One calorie equals 4.19 joules.

For the mathematically inclined or the just curious, appendix A at the end of the book gives a more detailed explanation of how the joule is calculated as well as examples and conversions to other energy measurements.

POWER—THE WATT

While many are not familiar with the joule, most of us have heard of the commonly used unit for power, the watt (named after James Watt, who invented the steam engine). Power means the amount of energy in a period of time. Specifically, a watt is equal to 1 joule per second. This is the globally accepted unit for power, although, as with energy, other units are used from time to time, some of which are also found in appendix A.

EXAMPLES OF POWER AND ENERGY—A WEE BIT OF MATH

1. Your lightbulbs, as well as other appliances, have a power rating printed somewhere on them. It will be in watts. If you have a 10-watt lightbulb, it uses 10 joules of energy for every second it is turned on. If you leave a 10-watt lightbulb on for five seconds, you will have used 50 joules of energy. Leave it on for five minutes (three hundred seconds), and you will have used 10 joules per second times three hundred seconds, or 3,000 joules of energy.
2. A 1,500-watt microwave will use 1,500 joules of energy every second. When you hit the thirty-second button on that microwave you will use 1,500 joules of energy per second for thirty seconds, or a total of 45,000 joules of energy to heat up your corn muffin.
3. You lift a bag of ten apples two meters from the floor in five seconds. The work you did is about 20 joules; therefore, your energy use is 20 joules. Divide this by five seconds and you have a power output of 4 watts (4 joules per second). Had you accomplished this same task in a single second, the work done would be the same, but your power output would have been 20 watts.

SOME COMMON PREFIXES/SHORTCUTS

Because a joule is a small amount of energy, and a watt is a small amount of power, we can get into some very big numbers quickly. Prefixes have been developed to keep us from getting lost in zeros.

Kilo = 1,000. Kilo means thousand. Put "kilo" in front of anything, and you have a thousand of them. A kilodollar would be a thousand dollars. A kilokangaroo, would be a thousand kangaroos.

Likewise:

- Mega = million
- Giga = billion
- Tera = trillion (A terakangaroo would be a trillion kangaroos, which is a serious kangaroo problem.)

Most of us recognize these standard prefixes as they pertain to computer memory:

- Kilobyte = 1 thousand bytes of memory
- Megabyte = 1 million bytes
- Gigabyte = 1 billion bytes
- Terabyte = 1 trillion bytes

The same prefixes are used to signify large amounts of power and energy:

- Kilowatt = 1 thousand watts = 1 kilojoule per second = 1 thousand joules/second
- Megawatt = 1 million watts = 1 megajoule per second = 1 million joules/second
- Gigawatt = 1 billion watts = 1 gigajoule per second = 1 billion joules/second
- Terawatt = 1 trillion watts = 1 terajoule per second = 1 trillion joules/second

When we are talking about energy that a power plant is capable of providing we often use the megawatt. A 500-megawatt power plant can supply 500 million joules per second to its customers.

We often express the amount of energy from the sun in terawatts. The sun provides the earth with about 700 terawatts of energy or about 700 million joules of energy per second. That is a lot of energy.

Examples of Using Prefixes

1. A power plant might provide 500 megawatts of power at full capacity. How much energy is this?

    - Mega = million. Therefore 500 megawatts = 500 million watts or 500,000,000 watts.
    - Since a watt is 1 joule per second, this power plant is capable of providing 500,000,000, or five hundred million, joules of energy every second it is running.

2. Your electric company likely charges you by the kilowatt-hour (kWh). It is easy to see the word "watt" and assume that is a measurement of power. It is not. Multiplying power by time (in this case an hour) gives us energy. How much energy is it?

    - A kilowatt is 1,000 watts, or 1,000 joules of energy every second.
    - A kilowatt-hour is the amount of energy used in an hour; 1,000 joules per second multiplied by 3,600 seconds in an hour is 3,600,000 joules of energy.
    - If you pay $.50 per kilowatt-hour, you are paying $.50 for 3,600,000 joules of energy. Seems like a bargain when you put it that way.

KEY TAKEAWAY POINTS

- The universal unit for measuring energy is the joule, which is a small amount of energy.
- The universal unit for measuring power is the watt, which is 1 joule per second.
- Prefixes are used to describe larger amounts of energy and/or power. Common ones are kilo (thousand), mega (million), giga (billion), and tera (trillion).

DISCUSSION QUESTIONS

1. Energy companies usually charge by the kilowatt-hour (kWh). Why do you think they use this unit rather than the joule?
2. On a sunny day, about 30 watts of solar power is provided by the sun per square foot. How many joules of energy would this be every second? How many joules in an hour? How many joules would shine on a square yard in an hour? (30 J, 108 kJ, 972 kJ)

Chapter Three

The Dance

Changing Forms of Energy

Science knows no country because knowledge belongs to humanity, and is the torch which illuminates the world.

—Louis Pasteur

In 2015 in the port city of Tianjin, in northern China, there was a series of explosions that left 173 dead and scores injured. What happened?

It was a basic energy conversion from one form to another. This kind of conversion goes on all the time. Every time you turn on your car's ignition, swing a baseball bat, or drop a pencil, you are making this conversion, albeit without the drama or destruction suffered in Tianjin.

There are a number of categories of energy, but all energy falls into one of two important basic classifications: *kinetic* energy and *potential* energy. The disaster at Tianjin was a rapid change from potential to kinetic energy.

Life, movement, and the world as we know it are a dance between these two types of energy. Sometimes it is an elegant, breathtaking dance, like a bird in flight or a well-played tennis return. Other times it can be a nightmarish disaster. But a dance it is.

The various forms, descriptions, types of energy, and the ease with which it dances within these categories can be confusing. Let's look at a few that are important as they relate to the earth's current energy troubles.

KINETIC ENERGY

If something is moving, it has kinetic energy.

The definition of energy is "the ability to make something move." Something has to be in motion to make something else move. When it is moving, it has the ability to make something else move; therefore, it has energy. We call that kinetic energy.

Consider a billiard table. When the balls on the table are all sitting there, you would not expect the red ball to begin moving all by itself. On the other hand, if the white ball rolls to the red ball and hits it, the red ball would move. The white ball had kinetic energy. It gave some or all of its energy to the red ball when it hit it and the red ball began to move.

The word "kinetic" refers to motion. An airplane is moving through the air; it has kinetic energy. Water is pouring over an embankment, kinetic energy again. The faster it is moving and the heavier it is (in science talk, the more mass it has), the more energy it has. But a thing must be in motion to move something else, and therefore it must be moving to have kinetic energy.

Some Types of Kinetic Energy

Anything in motion has kinetic energy. There are a few categories of moving things specifically relevant to climate change that need to be recognized.

Light, Otherwise Known as Electromagnetic Radiation

This is energy that is transported by electromagnetic waves and is basic to understanding our global warming situation.

These waves include the light we can see as well as a variety of waves that are not visible to the human eye, such as radio waves, microwaves, infrared waves, ultraviolet waves, X-rays, and gamma rays. All travel through space at the fastest speed in the universe. Nothing can travel faster than light. (Sorry, "warp speed" enthusiasts. Einstein's work demonstrates this, and until someone comes up with evidence to the contrary, that's our story and we are sticking to it.)

The speed of the wave vibration, or *frequency*, is what determines where on this spectrum it resides. Radio waves are very low frequency; gamma rays are the highest. Red visible light is a lower frequency than blue or violet. Ultraviolet rays, X-rays, and gamma rays carry a lot of energy and can be harmful.

The sun transfers energy to the earth via electromagnetic radiation. It sends waves of all these frequencies, but by far the most comes from infrared, visible, and ultraviolet (UV). Lucky for us, UV light is largely blocked by the ozone layer of the upper atmosphere.

Most light that reaches the earth is in the visible range. It leaves the earth in the infrared range.

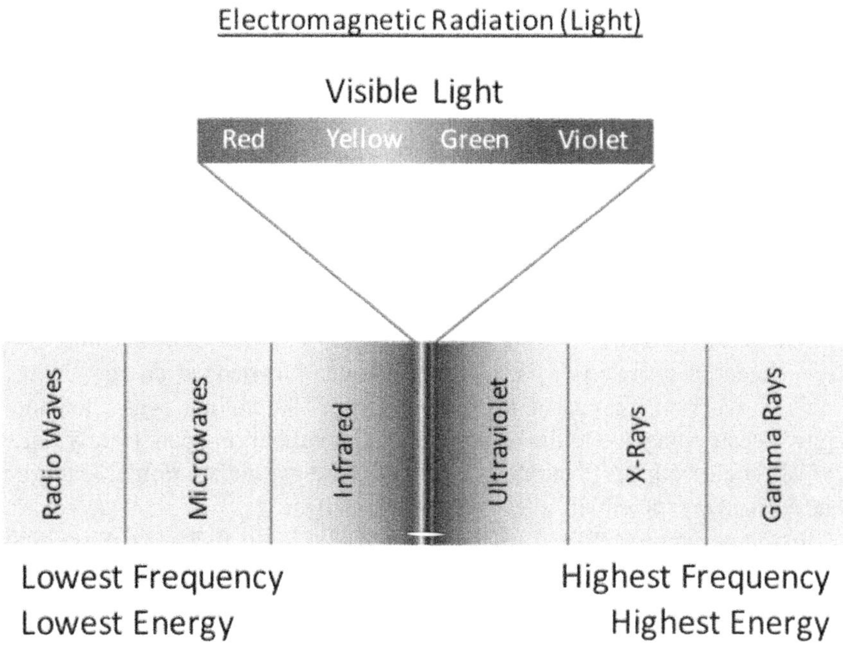

Figure 3.1. Electromagnetic Spectrum.

Heat

The motion of microscopic particles is felt as heat; more motion, more heat.

Electrical Energy

This energy is transported by moving electrons. This frequently travels through wires, but also includes a good lightning storm or the thrill you get when dragging your feet on certain types of carpet and then touching something metal.

Sound

Sound energy is also transported by waves, but in this case some kind of material is vibrating, it could be air, water, the ground, or train rails. These waves can set the ear's tympanic membrane in motion, which sends messages to the brain. We call this phenomenon "hearing." Very loud sounds can sometimes be "heard" (felt as vibrations) by other parts of our body. *Star Wars* was wrong. There is no way the explosion of the death star could be heard at the rebel base—sound waves do not travel through empty space.

POTENTIAL ENERGY

Stored energy is called potential energy.

This is energy that is in reserve, chilling out, waiting to be called into service. In its present state there is no motion, and it cannot do work, it cannot make something else move.

Why does it qualify as "energy"? Because it can be converted to kinetic energy, at which time it can do work. Potential energy can sit around for seconds, years, decades, or eons waiting for its moment of glory. But it is there, prepared for action. Potential energy does not do work until it is converted to kinetic energy, that is, until something starts moving.

The term "potential" may sound like it is a "possible" sort of thing. Not true. Potential energy is real, measurable, and full-fledged energy. It just can't do work in its present form. It is sort of like calling eggs "potential eggs" because they are in the refrigerator and not the frying pan. Hardly fair.

We could call this "waiting" or "stored" energy, but we don't. Tradition has decided on "potential" energy, so potential it is.

An important example of potential energy is fossil fuels—coal, oil, and natural gas. A piece of coal has been sitting in the ground for millions of years. Not moving. Not making anything else move. But under the right circumstances—say, inducement with a little fire—it turns into kinetic energy and can do work.

Similarly, the contents of the gas tank of your car are doing no work when the car is safely parked in a garage. But when a tiny bit of that gasoline is injected into a cylinder and then ignited—boom! The energy is released, hot gaseous particles move—and, in turn, move the piston. This is the brilliance of the "internal combustion" engine.

It is, of course, possible for an object to have both potential and kinetic energy at the same time. In fact it happens all the time. Your car cruising on the freeway at a cool 65 mph has a significant amount of kinetic energy based on that motion. It also is carrying potential energy in the gas tank. That potential energy is gradually converted to kinetic energy in the cylinders of the car. But in the tank, it is still potential.

Likewise, your body, that incarnation of motion, represents a lot of kinetic energy. It also is a repository for potential energy, a lot of which may reside around the hips. If anyone suggests that you may be a little chubby, it is perfectly appropriate that you point out you are merely storing some additional potential energy, just in case.

The smooth conversion of potential energy to kinetic energy allows the car or your body, or any number of other things, to move in a controlled fashion.

Some Types of Potential Energy

Types of potential energy take more explanation. But they are an important part of the global warming problem. In fact, we could call climate change a crisis of potential energy.

Chemical Potential Energy

This is energy held within molecules, which are atoms bound together. These molecules can be broken apart to release energy. The food you eat, wood you burn, gasoline in your gas tank, or the coal running a power plant are all chemical potential energy. Batteries, explosives, matches, and candles are also sources of chemical potential energy.

Fossil fuels are chemical potential energy that is easily stored and manipulated. Because of its "storability" it is convenient and used extensively. But it also means that we are using great quantities of chemical energy that have been stored for millions of years.

When chemical potential energy is turned into kinetic energy, a chemical reaction takes place that usually releases carbon dioxide. It is the release, in a relatively short time, of carbon dioxide that has been stored for millions of years that threatens us today.

The granddaddy of all chemical energy is floating around in your bloodstream at this moment. The sugar molecule (glucose) is an original carrier of chemical potential energy. We will discuss this more in the next section, but the potential energy in sugar is important to an understanding of fossil fuels and our climate change problems.

What happened that tragic day in Tianjin was an uncontrolled rapid conversion of large amounts of chemical potential energy to kinetic energy. Massive amounts of material began moving very rapidly in an extreme violent energy conversion, otherwise known as an explosion.

Gravitational Potential Energy (GPE)

This is energy that an object has by virtue of its place in a gravitational field (like Earth's). This can be a bit confusing.

If you lift a bowling ball and drop it on, say, a tomato, it will do work. Pieces of tomato fly around, juice all over the place, sound, maybe a dent in the floor. Things moved.

The higher you lift the ball, the more work it will do when it hits the floor. Bigger dent, louder noise, tomato bits fly farther, and so forth. That is because the higher the ball is, the more GPE it has. That ball gains potential energy the farther it is moved away from the earth. It was "given" GPE when someone did the work necessary to lift it. Motionless above the ground, its energy is "potential." When released, and the ball goes into motion, it now has kinetic energy with which it will do work on the tomato, and the floor.

Another important example. In early spring, look up into the mountains and what do you see? Some would say snow, others a future skiing trip, some a beautiful landscape. Any good scientist sees gravitational potential energy. All that frozen water contains a great deal of GPE that is released as the snow melts and the water begins to flow downhill to the ocean. If you have any doubts, take a look at Niagara Falls or raft down any decent white-water stream. That water has a lot of energy as it travels back to the sea.

The GPE of the snow becomes kinetic in the river and can be converted to electrical potential energy in hydroelectric power plants. This is an important, if limited, non-CO_2-producing form of energy.

Electrical Potential Energy

Electrical potential energy follows similar rules as gravitational potential energy, but it is more difficult to visualize. Electrons have potential energy based on their position within an electrical field. Nature creates electrical fields in the atmosphere that energize electrons and, given the right conditions, produces lightning. Humans create those fields in wires, through which electrons can carry and deliver energy.

Electrical energy is potential when it is sitting in the wires. But once you turn the hair dryer on, the electrons are in motion and we have kinetic energy. Electrical energy is extremely important both because of the fossil fuel energy that is converted in its production and the fact that it can offer some of the best solutions to our climate change situation.

Nuclear Potential Energy

Potential energy contained in the nuclei of atoms is nuclear potential energy. It is an enormous amount of energy. Like hydroelectricity, nuclear potential energy offers CO_2-free energy, but it does bring some problems. We will discuss the nucleus in detail in chapter 5.

One of the biggest challenges we face as we address global warming is how to convert solar kinetic energy into potential energy. How do we store some of the large quantities of solar energy coming to Earth every day?

THE DANCE—SOME EXAMPLES

1. Your body is calling on some potential energy reserves within the molecules of your cells. In highly controlled reactions, molecules are broken apart to provide the kinetic energy necessary to maintain bodi-

ly functions, keep your heart beating, dance a waltz, or push the buttons on the TV remote.

2. You have a rubber band. Hook one end over your left index finger and pull with your right hand. As you hold the rubber band stretched between your hands, it contains potential energy. No movement, but we know the energy is there. Let go with your right hand and that potential energy is converted to kinetic energy. The rubber band is on the move.
3. A rifle cartridge consists of a bullet encased with potential energy–filled gunpowder. Provide the right amount of friction to that powder, and the potential energy is rapidly converted to kinetic energy, sending the projectile on its way in an often dramatic, and potentially dangerous, fashion.
4. It happens the other way as well. We can convert kinetic energy into potential energy. Kinetic energy from rushing water can be converted to electrical potential energy using a turbine in a hydroelectric plant. That electrical potential energy can then be transferred to convenient locations and converted back to kinetic energy to run equipment, lights, gadgets, toasters, and telephones.
5. The kinetic energy from sunlight is being converted to chemical potential energy by the leaves of plants.

KEY TAKEAWAY POINTS

- There are two major classifications of energy: kinetic energy and potential energy.
- Kinetic energy is in motion, and can do work. Typical examples of kinetic energy are light, heat, electricity, sound, and any moving object.
- Potential energy refers to stored energy that has the potential to cause motion, but doesn't until it is converted to kinetic energy. Some important examples are chemical potential energy, gravitational potential energy (GPE), electrical potential energy, and nuclear potential energy.
- Kinetic energy and potential energy are routinely converted from one to the other.

DISCUSSION QUESTIONS

1. What do all forms of energy have in common?
2. Sound energy is called "kinetic energy," but what is moving?
3. What are some types of potential energy that you can see right now?

Chapter Four

Can't Be Made, Can't Be Destroyed

Just Keeps Dancing

The law of conservation of energy tells us we can't get something for nothing, but we refuse to believe it.

—Isaac Asimov

On February 20, 1962, a young, handsome test pilot named John Glenn strapped himself into a cramped space capsule installed on the top of a huge Atlas rocket that was filled with enormous quantities of rocket fuel (a.k.a. potential energy).

After running through the copious checklist that NASA is famous for, somebody pushed a button. With much sound and fury, the potential energy of the fuel was converted to kinetic, and Glenn was sent flying beyond the atmosphere, and thrust into orbit around the earth at a speed of about 17,000 mph.

For most of his three orbits, it was a quiet and peaceful voyage. He had no sense of motion despite his astronomical speed. He spent his time describing the remarkable view and thanking the people of Perth, Australia, for turning on the lights, which he could see through his small window.

Despite his practiced calm, Glenn knew the most dangerous part of his voyage—of any space voyage—was still ahead of him. We've all seen the simulated pictures as space capsules leave orbit in a fiery blaze that would seem to leave no hope for survival. Of all the thousands of things that must be accounted for in space travel, controlling energy in the reentry is the most crucial. Designing space vehicles to withstand this heat is a sobering engineering challenge. Even the smallest error can mean the difference between ticker tape parade and a deadly inferno.

Colonel Glenn's reentry was successful; he got the parade. The *Discovery* mission reentry in 2006, doomed by the loss of only a few heat-protective tiles, was not.

Tough as it is to travel into space, coming home is tougher still. What's going on here? Why all the pyrotechnics? Simply put, John Glenn's trial by fire as he reentered the atmosphere was a problem of conservation of energy.

CONSERVATION OF ENERGY

In the mid-1800s a young brewer with scraggly hair (the mark of a true scientist?) named James Prescott Joule used a clever apparatus to turn a paddle inside a water-filled, insulated container. The water got warm.

So what?

Science can be described as slow, tedious work punctuated by an occasional breathtaking discovery. This was one of the latter. By carefully measuring the work done to turn the paddle wheel, and the heat produced, Joule established one of the most important pillars of modern science called the "conservation of energy."

His experiment, and countless others since, demonstrated that energy is "conserved," that it is neither created nor destroyed but only converts from one form to another. In this case Joule proved that the energy used to turn the paddle against the resistance of water was not used up or lost, it was converted to heat.

Indeed, the amount of energy in the universe, as far as we can tell, remains constant. This is important to our global warming situation. Much as we wish we could, we cannot "use up" our way to a cooler earth.

To many this is counterintuitive. Certainly energy is "used up" in a car; the tank needs to be refilled on a regular basis. And what about power plants, aren't they making energy?

In both of these instances, energy is being changed in form, but not lost or made. In the case of your car, the energy gasoline is being converted into heat (while doing the work of moving the car), and dissipated into the atmosphere.

Power plants are converting energy into a more usable form, electricity. It is difficult to run a computer with coal, or power your refrigerator with falling water. The power plant is a middle man, but not a creator.

Even though energy changes methods of expression when doing work, it is all still energy, all measured in joules, and none of it is lost or made. Whether energy is used to boil an egg, kick a football, or heat a room, it is changing, not disappearing.

So how does this explain John Glenn's fiery return to Earth? How was that a conservation of energy problem?

Glenn's blazing rocket launch left him with a great deal of energy. Because of his enormous speed, he had more kinetic energy than a cruising 747. Likewise, his height above the earth meant he had a large amount of gravitational potential energy.

When he arrived back on Earth, he had neither the speed nor the height. Where had all that energy gone?

Conservation of energy says it doesn't just "go away." It must somehow be given away or converted to another form. In this case, once again, it was converted to heat—a great deal of heat—which was "given" to the molecules in the atmosphere. Hence, the fireworks on the way down. There was no avoiding this; the massive amount of energy he accumulated on the way up must be dispersed into heat, resulting in a space capsule encased in a very hot, glowing atmosphere.

Back to the bowling ball and tomato saga. When we left this story we had just made a mess. Tomato is done for, floor is dented, sound is over, and ball is back on the floor. Where is the energy? Just as with John Glenn's space capsule, an automobile, or your computer, it ends up heating the air around it, just a bit.

This is where Joule's discoveries become so important. The heat loss is rarely as dramatic as that experienced by Colonel Glenn, but it is there. And if you feel your car, or your computer when being used, or yourself for that matter, you will notice an off-putting of heat.

The other question is from where did the energy come that your muscles used to raise the bowling ball. That's easy—the chemical energy in that hamburger you had for lunch. In fact, the point of eating is to collect chemical potential energy either for immediate use or to be stored for future needs.

The hamburger (we will ignore the sesame seed bun, and special sauce for now) came from Edgar the steer, raised on a feedlot somewhere in Nebraska. It happened to be part of his right flank, and as a combination of protein and fat was chock full of chemical potential energy.

Turns out Edgar got this chemical energy from grass or possibly corn. How does the grass and corn get its chemical energy? From electromagnetic radiation (light) from the sun. Yes, back to the sun.

Bottom line: through an ongoing dance, the energy used by the bowling ball to smash the tomato was, originally, electromagnetic wave energy from the sun. Solar energy. And it eventually ends up as heat, the increased motion of molecules.

Heat is the final resting place for energy. It is dissipated in the atmosphere and eventually leaves the earth as infrared radiation—or most of it does.

The earth is routinely losing heat from our atmosphere. Energy comes in from the sun and leaves as infrared radiation. This cycle of energy in, energy out has been going on since the beginning of the earth and is essential for continued life on the planet. We will revisit that story when we get to chapter 11 on the greenhouse effect.

Energy that arrives on Earth as light (electromagnetic radiation) can be converted several times—smashing tomatoes, propelling cars, doing work on a variety of objects—and finally leaves as infrared radiation.

So, energy ends up dissipating into space. But where does it get its start?

KEY TAKEAWAY POINTS

- Energy is neither created nor destroyed. To our best understanding, the amount of energy in the universe is constant.
- Energy is continually being converted from one form to another.
- When energy "does work" it is undergoing a conversion; the amount of energy remains the same.
- Almost all our energy comes from the sun and electromagnetic radiation (light). It goes through conversions, does work, and ultimately returns to space in the form of infrared radiation.

DISCUSSION QUESTIONS

1. The "sesame seed bun and the special sauce" part of the hamburger, as well as the lettuce, onion, and tomato were ignored. What kind of energy do they have? Where did it come from?
2. Snow in the mountains has been mentioned as a great source of GPE. Explain how the snow got that potential energy, and where it goes as the water flows back to the sea (assuming no power plant intervenes).

Chapter Five

The Wellspring of all Energy

Surprise! It's All Nuclear

Protons give an atom its identity, electrons its personality.
—Bill Bryson, *A Short History of Nearly Everything*

If energy is not created, from where does it come originally? It may surprise some people, but all the energy in the universe was, originally, nuclear potential energy. This will take some explanation and a smidgen of physics. But, it is doable, so hang on.

PROTONS, NEUTRONS, AND ELECTRONS, OH MY

We are surrounded by matter: everything we are, see, feel, breathe, or eat. But what is matter, and how many kinds are there?

Originally, the Greeks thought everything was made from earth, water, air, and fire. Over the last two hundred years or so, some very clever people have determined that everything—from fingernails, to plastic dinosaurs, to cruise ships—are made of various combinations of ninety-two unique types of atoms (referred to as the "elements"). They are conveniently lined up in an awkward-looking "periodic table of the elements" that appears in classrooms and on T-shirts around the world.

Some will note that there are actually more than ninety-two elements on that chart, but those beyond ninety-two do not occur naturally, they are man made, and generally hide away in laboratories rather than our public spaces.

Plutonium

There is one man-made element that exists in significant quantities. That is plutonium, number 94. It is used for making bombs—and is a product of some nuclear power plants. Today there are about five hundred tons of this material on the planet. It is dangerous and difficult to safely store. Nobody likes the stuff, so we will ignore it—for now.

Each element is made of different atoms that have widely different characteristics and behaviors. But all atoms—from number 1, hydrogen, a very light gas, to number 92, uranium, a very heavy metal—have important things in common.

All atoms are made of different numbers of three smaller particles called protons, neutrons, and electrons. They all have a nucleus in the center, where the protons and neutrons hang out, and a cloud of electrons in motion around it. You've seen diagrams of atoms: a bumpy group of marbles in the center and a few things orbiting around the outside. Cute, but misleading.

The nucleus is tiny. Beyond tiny. If the nucleus was a fly on second base at Fenway Park, the closest electrons are in the last row of the nosebleed seats. Not into baseball? If the nucleus was the size of an orange, the atom would be the size of Manhattan. Admittedly, Manhattan is an awkward shape for an atom, but for purposes of size comparisons, it gives us a rough estimate.

Atoms are almost entirely empty space. Things feel solid because of the electric/magnetic forces within the atom, sort of like how a balloon feels solid. But our world is not as solid as it feels.

The nucleus, residing deep within the atom, contains virtually all the mass of the atom. Further, it decides just what kind of atom it is. The number of protons in the nucleus is the number of the element, and determines what the element is, how many electrons it has, and how it behaves. If you have 3 protons in your nucleus, you are lithium—no arguments, no negotiations. If you have 6, you are carbon. Thirty-three you are arsenic.

The other members of this nuclear family are the neutrons. Their number can vary, sometimes similar to the number of protons, sometimes varying widely. Neutrons have other functions, including stability. But it is protons who rule.

Protons attract electrons, and the number of electrons in a neutral atom is the same as the number of protons. Therefore, each element has a different number of electrons in its cloud shell. And it is here, based on number and arrangement of electrons in the cloud, where all the action takes place.

The electrons are the social part of the system; they determine the activity of the atom. Some elements are total snobs—helium and its whole snobby family, for instance. They have nothing to do with anyone, not even other helium atoms. Electrons in these elements stick close to home and ignore the world. They are extreme introverts.

Sodium, at the other extreme, is a poster child of extroversion. It reacts explosively given the opportunity. Sodium electrons are only too happy to mix it up with others. Other elements, say, carbon and its family and neighbors, are more measured, but definitely social creatures, which is why carbon is so useful in biological systems.

The whirr and excitement in the atomic world is all the action of the electrons that inhabit the outer reaches of the atom. Electrons sometimes move from atom to atom, leaving home to bond elsewhere or just fly around for a while. Molecules, which are atoms that join together in a group, do so at the behest of electrons looking for a compatible arrangement. Chemistry is the study of how the atoms of various elements react with each other. And that is all the business of the electrons.

The tiny nucleus resides deep inside, like a powerful but detached dictator, unaffected by the hysterical groupings and societal activities of the electrons. Unfazed, untouchable, unassailable, peacefully unchanging, forever and ever.

Or so we thought.

Turns out life is not necessarily all that peaceful in the sovereign nuclear world.

NUCLEAR REACTIONS

Until the 1930s it was assumed that atoms could not "transmute," that is, they could not change the number of protons in the nucleus and become a different element. The nuclei of the elements in our world were forged in the stars, and it was assumed that there was no way those nuclei could change or rearrange. If you were a silicon or uranium atom, you would remain forever silicon or uranium.

We now know that nuclear changes and transmutations can and do take place. All the time.

Keep in mind, however, that with the exception of a tiny amount of radioactivity, this does not happen in your body, your kitchen, your garden, your car, your rocket, or even in your head after smoking something inappropriate. Nuclear reactions are largely star-stuff, what makes the stars twinkle. They also occur deep inside planets like our own.

All nuclear processes convert a great deal of nuclear potential energy to kinetic energy, which is why they are important to us.

NUCLEAR POTENTIAL ENERGY VERSUS CHEMICAL POTENTIAL ENERGY

Chemical reactions are reactions among atoms that come together in groups forming molecules, rearrange into different molecules, or sometimes going freelance. Chemical potential energy can be released when the atoms are rearranged into more stable configurations. This is what runs our cars and bodies.

Nuclear potential energy is that energy contained deep within the nucleus inside of atoms and is released when the *nucleus* is changed into a more stable configuration.

Nuclear reactions yield far more energy per mass than chemical reactions. One comparison would be the conventional "chemical energy" bombs commonly used in warfare, compared to nuclear bombs. Large conventional bombs can destroy a building. A nuclear bomb, as we well know, can flatten a city. The amount of energy tied up inside the nucleus is astounding.

To a humanity with constantly increasing energy wants, the staggering amount of potential energy within the nucleus can seem like a panacea. One of the basic problems is how to use it in a practical way. To that end, we need to look at each of the ways that nuclear energy is released both by nature and by humans. There are three basic types of nuclear reactions:

1. *Radioactive decay*—The breaking down or rearranging of unstable nuclei into more stable configurations. This occurs naturally.
2. *Nuclear fission*—A large, heavy nucleus breaks apart into smaller ones as a result of a bombardment and resulting chain reaction. This is a process developed by humans.
3. *Nuclear fusion*—Two small nuclei join together to become a heavier one. This is the reaction that runs the stars.

RADIOACTIVITY—KEEPING THE INSIDE OF THE PLANET WARM

The first hint that there was action in the nucleus occurred in 1895 when a French physicist named Henri Becquerel noticed that certain materials emitted a type of radiation that could be picked up on a photographic plate. It was both shocking and exciting. Marie Curie coined the term "radioactivity" and established that the radiation was coming from the nucleus of the atom. Eventually three forms of radioactivity, alpha, beta, and gamma, were identified, as well as an explanation of what was going on in the nucleus.

As natural atoms get bigger, the nucleus is loaded with more and more protons and neutrons. The nuclei are packed. Relative to the whole atom, the nucleus is still ridiculously small, but down in nucleus-land it is getting really big and cramped.

The protons all have a positive charge. "Like" charges repel so it gets pretty stressful. One reason for the increased number of neutrons in these heavy elements is that they offer stabilizing forces to hold these protons together. But a great deal of energy is required, and eventually, even the neutrons can be overwhelmed.

It's a little like the holidays. A big, and not necessarily congenial, family has gathered itself into a very small house. Tensions abound. People are gritting their teeth and its getting hard to hold the festivities together. Finally, things come to a head and Uncle Joe, Aunt Susie, and their obnoxious children are ejected from the gathering. There is a sigh of relief, the fruitcake is passed around, and an uneasy calmness prevails. At least for a while.

In radioactivity this "family" is an unstable nucleus, perhaps uranium, with protons and neutrons packed in. A grouping of two protons and two neutrons, a helium nucleus, is dramatically expelled from the group. This helium nucleus goes on to pick up a couple of electrons—not a difficult feat, electrons are often out looking for new relationships. A helium atom is formed and on the loose.

Meanwhile, the atom left behind has two fewer protons; therefore it is no longer uranium (92 protons)—it has become a different element, thorium (90 protons). Transmutation has occurred.

Before we knew this expelled item was a helium nucleus, it was called an "alpha particle" and it is still called "alpha radioactive decay."

There are two other types of radioactive decay: beta and gamma. Similarly, they eject a piece of the nucleus to reach a more stable configuration, usually resulting in a change of element. These "ejections" are picked up as "radioactivity," which can be heard as beeping sounds on a Geiger counter. Gamma radiation is the kind that is dangerous in large amounts because penetrating high-energy photons are emitted.

Worth noting is that there are also a number of lighter elements that, for one reason or another, have nuclei that are unstable and also demonstrate radioactivity. One of the more famous is carbon-14. But it is the heavy guys that are the real champs.

What is important for our purposes, however, is that the energy that was being used to hold this dysfunctional group together is no longer needed. It is sent out into the world as kinetic energy, moving things around, heating things up. And, it is a lot of energy. Nuclear reactions release a great deal of energy.

This process is going on right now deep within the earth, and has been since the earth was first formed about 4.5 billion years ago.

Energy from radioactivity is what fuels the great molten center of the planet. The plates of the earth are under constant pressure from this nuclear heat and molten rock flow. Potential energy in the tectonic plates builds up until, finally, it is released and causes massive movements of the earth's crust, known as earthquakes. Breaks in the crust lead to volcanoes. The pressure builds from below until there is an enormous outpouring of kinetic energy, lava, gas, and ash.

Geysers, boiling mud pots, tar pits, steam vents, and other fascinating and sometimes scary phenomena are all a result of the energy released by ongoing nuclear radioactivity within our earth. So is the helium that fills those confounded balloons and makes your voice sound funny. Helium is too light to stay in the earth's atmosphere so, like infrared radiation, it is something we are leaving behind as we make our way around the sun.

Eventually our earth will run out of these naturally unstable atoms and radioactivity will come to a stop. But people who know about these things say we are good for a couple of billion more years so not to worry for the time being. The joy of volcanoes and the excitement of earthquakes are pretty much here for the duration. These sources of energy have been important to life on earth, originally providing us with an atmosphere and causing crust movement that has had effects on the origins and evolution of life.

While we have this tremendous source of energy beneath us, much of it has remained difficult to contain for human purposes. Controlling a volcano or getting usable energy out of an earthquake pretty much eludes us. Hot springs, sometime steam vents, have been put to good use, and in certain areas of the world the heat can be converted to electricity or heat for homes and buildings. This heat within the earth is seen as a viable and nonpolluting, if limited, form of energy.

NUCLEAR FISSION—A DESPERATE INVENTION

In the 1940s, we learned how to release energy from the nucleus of certain heavy atoms in a process called "fission." The amount of kinetic energy released from the nucleus by this means is astronomical.

The discovery of nuclear fission came out of desperation. We had reason to believe that the Germans were close to figuring out how to release energy from the nucleus and develop a new type of catastrophic weapon against which there was no defense. The feeling was that the only hope for Europe and the United States was to beat them to it. As it turned out, they were not even close, but the United States continued with the research and created new and terrible weapons that changed the world forever.

Simply stated, by bombarding enriched uranium or plutonium (an unstable man-made element) with neutrons, we can set off a chain reaction that breaks these very large elements into two (sometimes three or more) midsize elements, converting pent-up nuclear potential energy into kinetic energy.

The initial use we found for this marvelous invention was the two bombs dropped on Hiroshima and Nagasaki to end World War II. Currently, we store thousands of much larger bombs.

We have learned, since then, how to design power plants where fission reactions can take place in a controlled environment and can be started and stopped at will, pretty much. We also power warships with controlled fission reactions, which has changed the nature of naval warfare.

Nuclear fission power plants, unlike the burning of fossil fuels, produce a great deal of energy without releasing CO_2 into the environment. This energy is converted to electrical energy. There are drawbacks, which we will discuss in chapter 21. But, despite its problems, many see it as part of a short-term solution to our energy needs while we develop sustainable solar energy technology.

NUCLEAR FUSION—TANTALIZING AND ELUSIVE

Fusion is the energy of the stars. It is what we see twinkling in the night sky. Thermonuclear bombs, more powerful by orders of magnitude than their fission cousins, use fusion (as well as fission to set the reaction off). Fusion is a naturally occurring process that works by a completely different process.

In this case, two hydrogen nuclei (containing one proton and one neutron each) are fused together and become a helium nucleus. Helium is a very stable atom, and the two hydrogen atoms give off energy when forced together. It is a marriage of sorts, a very compatible one.

At the beginning of the universe, the first and only element was hydrogen, which still makes up about 70 percent of the mass of the universe. The hydrogen has been fusing it into helium ever since, giving off astronomical amounts of kinetic energy in the form of electromagnetic radiation.

Heavier elements have resulted from further fusion in the massive pressures that occur when stars die, supernovas and the like. You, being made largely of heavier elements, are, quite literally, made of the products of fusion in these great explosions—stardust.

In many ways, fusion would be the ideal solution to the increasing energy needs of the planet. It is clean. It does not produce radioactive waste products, just calm, inert helium. The raw materials are readily available with a minimum of processing; the hydrogen would come from water, a substance of which we have, for practical purposes, an endless supply. There is no chain reaction so the likelihood of the reaction getting out of control is less.

The main problem is that both tremendous heat and pressure are required to get the reaction started. And containing the heat is problematic with the required temperatures melting any substance. This is, after all, the power of the stars where unimaginable gravitational forces provide pressure and heat.

So far our only significant success in releasing energy through fusion has been through enormous bombs—that is no way to cook dinner.

Extensive, costly efforts are under way around the world to find ways to harness this remarkable energy source for peaceful purposes. Thus far, any energy resulting from artificial fusion has generally been a great deal less than the energy required to produce it. Success has eluded us and is unlikely in the foreseeable future, if at all.

The standard joke in the industry is that we are about thirty years away from harnessing fusion, and have been for the last sixty years. Since we cannot wait thirty years to replace fossil fuels with clean energy, earthbound fusion is not seen as a solution to our current global warming problem.

And yet, the idea is so tantalizing. Fusion is the holy grail of energy research. The energy from fusion could provide vast quantities of clean energy to fuel the foreseeable needs of humanity well into the future.

If only we had a big, safe fusion reactor.

Turns out, we do.

KEY TAKEAWAY POINTS

- The source of all energy is nuclear potential energy, from within the nucleus of the atom.
- There are three types of energy producing nuclear reactions:
    - Radioactivity, the natural decay of unstable atoms
    - Fission, the splitting of the large unstable nuclei into smaller atoms
    - Fusion, the combining of two hydrogen nuclei into hydrogen
- Fusion is the energy of the stars, and the energy of our sun. It releases vast quantities of clean energy from readily available materials, but so far we have been unable to control the reaction in a practical way to provide for our energy needs.

DISCUSSION QUESTIONS

1. What are some ways you could see that the earth's internal heat could be used for human purposes in homes, industry, or transportation?
2. Nuclear fission reactors can produce a great deal of electricity without the CO_2 emissions of fossil fuels. There are, however, some drawbacks. What is your general feeling about the use of nuclear fuels?

Part II

The Ways of the Natural World

Chapter Six

Homage to Sol

Even in the midst of the storm the sun is still shining.

—Dayna Lovely

About two miles from Asbury in southern England stands a prehistoric structure we now call Stonehenge. It was constructed in about 2500 BCE from dozens of massive stones weighing an average of twenty-five tons, and transported, using muscle and determination, from twenty to two hundred miles to their current location. This was no afternoon project. Estimates are that the building of this sacred monument took about fifteen hundred years; some suggest it was under construction for as long as six thousand years.

The builders left no writing so we don't know exactly what activities went on within and around it. But we do know that the stones line up perfectly with both sunrise at the summer solstice, the longest day of the year, and sunset at the winter solstice, the shortest day of the year.

Whatever else this cathedral was, it was paying homage to the sun.

Stonehenge is only one (perhaps the most famous) of dozens of structures and artifacts we have from ancient peoples that are oriented around the comings and goings of the sun. It is a theme that occurs repeatedly in antiquity.

Our distant forebears, from vastly different cultures and locations, have worshipped the sun as a god as far back as we have records. It has gone by the name of "Amaterasu" (Japanese), "Apollo" (Greek), "Garuda" (Hindu), "Inti" (Inca), "Lugh" (Celtic), "Sol" (Norse), "Ra" (Egyptian), and many others.

Is it possible these ancient peoples comprehended something that has been largely lost on contemporary humanity? At a deep level, they understood that the sun was the source of continued life and everything they held dear.

The sun is just as important to us, but in all our sophistication we can lose track of the life-giving forces around us and how dependent we are on them. It is time to develop the type of reverence our ancestors had for the workings of our favorite star, with a modern understanding.

WHAT IS THE SUN?

The sun is a huge nuclear fusion reactor putting out unimaginable quantities of energy, a small percentage of which falls on the earth. That little bit is enough to meet all our needs with plenty left over to fulfill whatever wants future generations may deem necessary. There is more sun energy striking the earth in one and a half hours than worldwide human consumption for a year.

To be more precise, the sun provides about 700 terawatts of energy. That's 700 trillion joules every second. Bottom line—it's a lot of energy. This has been going on for four billion years and, we are told, it's good for four billion more. All of humanity currently uses about 17 terawatts.

Solar energy runs the world, in a plethora of ways. If we are to envision a world that remains livable for future generations, we need to understand how this is accomplished and how we can work within these natural processes rather than fight them. We will need an adjustment in some of our basic assumptions.

WHAT GOES 'ROUND, COMES 'ROUND

As a rule, people think linearly. We move from place A to place B. Whether it is a plastic bag, a pair of jeans, or a rocket ship, we make something, use it, and throw it away. It's done. We see "progress" as a single direction.

Nature, however, doesn't think that way. It goes around in circles. Old stars recycle into new stars, seasons reliably move through their yearly rotation, the earth turns, death provides materials to be used again for new life. Nature does not throw things away; it reuses over, and over, and over again. Nature is a series of cycles.

Science has clarified how the many great natural cycles work. There are a lot of them and they are interrelated and interdependent, but we describe them separately for clarity. Their names are remarkably uninspired: the "rock cycle," the "water cycle," the "carbon cycle," the "nitrogen cycle," the "sulfur cycle," the "cycle of life and death." But these cycles make the difference between a living planet, such as our own, and a dead one, such as the moon.

And they all depend on a constant supply of energy from the sun, in the form of electromagnetic radiation.

Our most serious environmental problems, including global warming, stem from a lack of understanding of these cycles, and of how our activities can interfere with them. The oft-used word "sustainability" means nothing more than working within these natural systems. It requires us to move out of our linear thinking and into the mind-set of nature that has nurtured life, including ours, for so long.

Two of these great life-giving, sun-powered, cycles are the main players in our global warming drama—the water cycle and the carbon cycle. We need to spend some time getting to know them.

KEY TAKEAWAY POINTS

- Our ancient ancestors had a deep understanding of the importance of the sun to life, an understanding that, while still just as true, our modern world lacks.
- All of nature and life depends on great sun-driven cycles. Understanding these cycles is key to understanding how the activities of humanity are affecting them, and how we can restore balance to the natural world.

DISCUSSION QUESTIONS

1. Why would our ancestors of three thousand years ago have more of a sense of the importance of the sun to their everyday lives than we have today?
2. What is it about the winter solstice (the day with the least sunlight) that would cause fear, celebration, and worship by ancient people?
3. Nature works in cycles. What are some things in your life that are cyclic? What things do you think of as linear (beginning, middle, end)?

Chapter Seven

Storms Brewing

The Water Cycle

Rain is grace: rain is the sky descending to the earth; without rain, there would be no life.

—John Updike

In *The Wizard of Oz*, little Dorothy, Toto the dog, and her entire farmhouse are whisked away into the far reaches of the atmosphere, ultimately landing, relatively intact, on a witch. While the unscathed landing in a fantasyland where scarecrows sing and tin men dance stretches our scientific credibility a tad, there is no denying that the power of an extreme storm can destroy or move a building, a tree, a car, or the contents of a garbage can across several counties. Anyone who has lived through a hurricane or a good old-fashioned thunderstorm appreciates the power in such events.

What is happening? All this turmoil is due to the relentless cycling of water within our gaseous atmosphere and it has been going on for billions of years all around the planet. Our sun is to be blamed—or, more appropriately, revered—for all this excitement.

The oceans, covering 72 percent of our planet, serve as a vast repository for water. Sunlight provides the energy to heat that water, as well as the air and the rest of the planet. It all follows from there.

Initially the water cycle seems relatively simple. The top layers of ocean water collect enough energy from the sun to warm and "change phase," turning from liquid into a gas. This is evaporation and is the same thing that happens when you hang out your laundry or boil water for spaghetti.

Warm water molecules, along with the warm air, rise in the atmosphere, where they cool again and gather as tiny droplets in clouds. Much of the water in these clouds, still propelled by solar energy, moves over the land where it falls as either liquid water (rain), or frozen as hail or snow.

Once back on earth, the water gathers together into lakes, rivers, and streams, while some of it seeps into the ground to make aquifers. The water in the streams and rivers moves downward toward the sea, carving the landscape of the planet, the Grand Canyon being a dramatic example of this.

Snow in the mountains, again using solar energy, gradually melts, keeping streams and rivers running throughout the warmer parts of the year. In some cold areas, the sun's heat cannot keep up with the accumulation of snow and it is never totally melted away. Every year a little more collects. This is how glaciers are formed and why the poles of the earth are permanently covered with snow and ice.

Ultimately, the water that was once in the ocean returns to go through the cycle again.

THE GREAT DISTILLERY

The first thing we gain from this process is a constant supply of freshwater. Seawater is, of course, salt water. Land animals such as ourselves cannot survive on salt water. But when the water evaporates from the ocean it leaves the salt behind. This process is called "distillation," or, in this case, "desalinization."

All rivers, big and small, all lakes, all aquifers are part of this unending cycle, and all water was once in the sea and will be again. Freshwater actually makes up a small proportion of the water on the planet. But all of it comes from, and is constantly replenished by, the water cycle, which is run by energy from the sun.

COMPLEXITIES

Of course on a large planet like ours, there are a great number of factors that complicate this cycle. The earth and the oceans do not warm evenly. Shallower water, different amounts of vegetation, assorted topographical features, dark colors, light colors, and cloud cover affect how much heat is absorbed and the temperatures of air and water in various places.

Warmer air and water rises. Water and air moves from places of higher pressure to lower pressure, all over the planet. This "movement" can be a gentle breeze, a wind, or a hurricane.

Because of the tilt of the earth we have seasons, which means that the amount of sunlight at any one place varies throughout the year. Different areas heat and cool at different times. Add to this the earth's rotation, causing a daily cycle of heating the land and the seas, and we have an ocean and an atmosphere that are dynamic. Great cycles and streams of warm air and water rise only to cool and fall, and be warmed again.

This, in short, brings about the weather. Wind, clouds, rain, sleet, hail, snow, scorching heat, all weather is essentially a result of the energy of the sun heating and evaporating water and turning the fluids of the world into continual energy-filled action.

The topography of the earth, the size and placement of the oceans, and the amount of sunlight reaching the earth tend to be relatively stable. Consequently, weather follows certain patterns. It snows in certain places, rains in others, and can be hot and bone dry in still others.

Weather is what happens day to day. The average of the weather, taken over a period of thirty years or so, is considered the climate. Climate is more dependable than weather.

Rain in a California summer is unusual; monsoons in an Indian summer are to be expected. Snow in the Rocky Mountains during January is the norm, but a blizzard in January in the Blue Mountains of Australia would be astounding. These rhythms and patterns are what make up climate.

All life, including humans, adapts to and depends on these climates. People in the tropics don't usually have parkas in their wardrobes, while those in the Russian tundra have very few tank tops. A bear hibernates in winter. Many animals typically have their young in the spring. Trees lose their leaves in autumn in preparation for winter cold and storms. Farmers plant their corn in spring, take advantage of summer rains, and harvest it in autumn. Certain crops are grown in certain areas because of climate. Certain animals thrive in certain climates and fail in others. Diseases thrive in some parts of the world and are absent from others because of climate. We largely take these climate patterns for granted.

If global factors change, climate can change. As the atmosphere warms up, more water evaporates, causing drought in certain areas. But that additional moisture in the air results in more and stronger storms and extreme weather events. Changing temperatures of the atmosphere change the "normal" airflow, changing the weather—and climate—around the globe.

This leads us to the other great cycle, closely intertwined with water and weather, known as the carbon cycle.

KEY TAKEAWAY POINTS

- The water cycle, run by the energy from the sun, provides a constant supply of freshwater to the land, including storage in the form of snow in the mountains.
- It also runs the weather, storms, and wind, in a reasonably predictable manner, otherwise known as climate.

DISCUSSION QUESTIONS

1. If water did not evaporate from the ocean, how would that affect life on land?
2. How does life adapt to hot, humid environments? What adaptive responses are needed in dry, desertlike environments? What about environments that are very cold most of the year?

Chapter Eight

Solid Sunshine

The Carbon Cycle

> *Imagine a machine that uses solar energy to remove carbon from the air and turn it into a beautiful strong and sustainable building material.*
>
> —Hippie Peace Freaks

The most enormous living thing on the face of the earth is the General Sherman, a redwood tree in Northern California. At two thousand tons, give or take, it beats out the largest whales and dwarfs a family of elephants. Ironically, this well-named *Sequoiadendron giganteum* comes from a seed roughly one-eighth of an inch in diameter. Given its humble beginnings, where did all those tons and tons of tree come from?

One might surmise that the soil around it was somehow converted into wood. But then there would be a gaping hole, or at least a bit of an indentation where soil was taken up. Awkwardly, there is none.

Indeed, if you look carefully you will see just the opposite. Any tree will do; you don't need to travel to Northern California. Just step outside and check one out. In a pinch, a sizable bush will work. The dirt around a plant is higher than the surrounding area. The tree has developed an underground root system, often big and powerful enough to raise sidewalks or do mischief on the occasional building.

Billions of years ago, our planet was all rock; plants have broken down that rock and added their decomposed bodies to produce soil. They don't take stuff away from the dirt—they add.

So, if not from soil, from where did that tree come? It is not too far off the mark to say it came from thin air.

Plant a seed. Any seed. Perhaps a bean, a tomato if you prefer. Put it in the ground, in a sunny location, and sprinkle some water on it. Remarkable things will happen. Spurred on by the moisture, roots will emerge and travel downward, seeking still more water. Then a stem will appear and travel the other direction, sending one or two leaves beyond the dirt surface. The leaves will turn green.

By any standard, this is a miracle, a fact lost on us because of its commonness, but a staggering miracle nonetheless. Our lives literally depend on this process—on those little green leaves.

Leaves are nature's solar panels. Their green color is a signal that they have begun one of the most astonishing chemical processes of the universe. This tiny plant is collecting sunlight and storing its energy in molecules that can be used for both construction and energy storage, not just for the plant, but for all living creatures. In fact, this is the only source of energy for essentially all life on the planet.

Life without Sunshine

There are some bacteria deep in the oceans that live off geothermal heat or energy from ionizing metal, rather than sunlight. These are important from an evolutionary standpoint, but as a practical matter, especially in the study of global warming, we can ignore them for the time being.

This process is called "photosynthesis," a word that comes from two Greek roots. "Photo" means light, and "synthesis" refers to making or building. Hence, photosynthesis means "making from light," which is exactly what is going on. The soil serves as a sponge to hold water, provide stability, and store tiny amounts of minerals. But it is the air, rain, and sunshine that plants use to build themselves and collect energy for all life.

The plant (the word "plant" in this book will refer to the huge variety of photosynthesizers, protists, bacteria, etc., that inhabit our world and collect light energy) gathers water and pulls in carbon dioxide from the air while collecting energy from sunlight to build a remarkable little molecule called—believe it or not—sugar. Specifically, this sugar is called "glucose" and it is the same stuff you have coursing through your body and feeding your brain right now.

The General Sherman, like all trees and all plants, from roses to seaweed, is made primarily of sugar. Sugar is the linchpin of life; it is our link to the sun and without it this world be a bleak, lifeless, planet.

Please do not choose to celebrate this fact by downing a twenty-four-ounce Big Gulp at your local 7-Eleven. What we have done commercially to this sweet little molecule is fodder for another book. But it is not possible to overstate the importance of sugar for the world we live in.

Understanding the process of photosynthesis, the creation of sugar, and what sugar is used for will help us understand the basics of today's global warming crisis.

CHEMISTRY

Some find the ideas of chemistry exciting, fun, and challenging. Others may think of a chemistry class as a cross between the Inquisition and the joy of a two-hour piano recital for five-year-olds. Out of respect for the latter group, this will be a brief toe-dipping into the subject with only one chemical equation. (Well, actually, two, but the second is the reverse of the first—a cycle.) And, it's important. So hang on.

Chemical reactions are pretty much like recipes, just written in code. Both give a list of ingredients and a product—either a gastronomical triumph or some sort of a new substance.

The recipe for sugar reads something like this:

1. Combine 6 molecules of carbon dioxide with 6 molecules of water.
2. Add energy from the sun.
3. Stir.
4. Yield will be 1 sugar molecule and 6 molecules of oxygen.

Bon Appetit!

A chemist will write the same thing as follows:

$$6\ CO_2 + 6\ H_2O + \text{energy from the sun} \rightarrow C_6H_{12}O_6 + 6\ O_2$$

Figure 8.1 on the next page is a diagram of this process in English.

The arrow in the chemical reaction (or "stir" in the recipe) represents a staggering series of complicated biochemical steps performed by every leaf on the planet; fortunately, not something we need to address here. The important thing is to realize that plants take carbon dioxide and rainwater and, using solar energy, produce glucose that is both an all-purpose building material and a storage cell for energy.

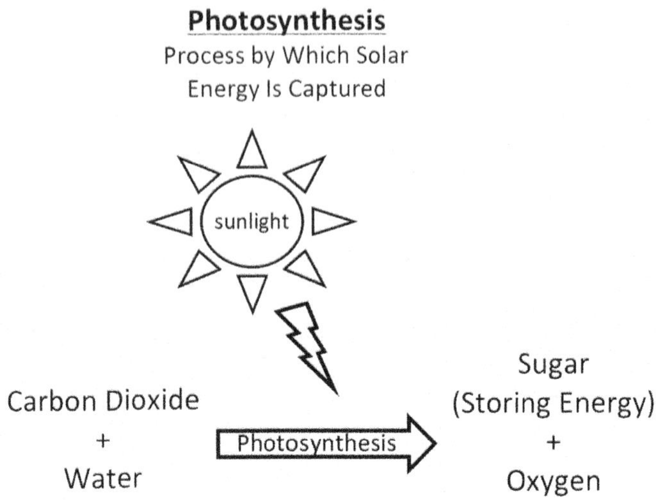

Figure 8.1. Photosynthesis.

There is one other substance at the end of the equation: O_2. This is oxygen, one of the gases you are breathing right now. Oxygen in this equation is like the eggshells or orange peels in a typical recipe. It is waste. In this case it is expelled into the atmosphere.

When plants first learned to photosynthesize, billions of years ago, the atmosphere had little oxygen. The oxygen in Earth's atmosphere has come primarily from the photosynthetic process of plants. We animals are still dependent upon them for a continued supply of the stuff. That is why the air seems so fresh in a place of dense vegetation like a forest. In addition to constructing trillions upon trillions of glucose molecules, the plants are pumping out vast amounts of oxygen every second.

Sugar by Any Other Name

Looking at the equation, you might notice that carbon is being added to water with the "dioxide," or O_2 expelled as waste. When water is added to something, it is "hydrated." Therefore, the plant "hydrates" carbon to make sugar. This is why we call sugar, in its various forms, "carbon-hydrate" or, more commonly, "carbohydrate." There are many other carbohydrates derived from glucose (sucrose, fructose, cellulose, amylose, etc.). A rule of thumb is that if you see the letters "ose" at the end of a chemical name, it is usually referring to some form of carbohydrate, which is some form of sugar.

GLUCOSE: LIFE'S UNIVERSAL BUILDING BLOCK AND ENERGY STORAGE

Okay, plants make sugar—a lot of sugar. Tons and tons of it every second. How do we get from "sweet" to "redwood tree"? And what else do we do with all this sugar? Three words: just about everything.

Much of the world is constructed largely of this little sugar molecule. Plants string long chains of it together into a tough substance called "cellulose." This fibrous material makes up the bulk of most plants, including a redwood tree. Anything constructed from plant material—your wood coffee table, the cotton curtains on the window, the pages of a book or magazine—are also made of cellulose, or chains of sugar.

Don't bother to chew on your kitchen cabinets. They won't taste sweet and you could get splinters. You don't have the enzymes needed to break the bonds between glucose molecules. Plant bodies can provide fiber, cellulose, in your diet. Many bacteria have the capacity to break the bonds of cellulose; hence, wood and such materials can be decomposed. But not by you.

In addition, glucose is the basis of all the various chemicals and chemical tools needed for both plant and animal life. There are literally thousands of other biochemicals derived from glucose. Other elements including nitrogen, sulfur, and phosphorus are added to it, and rearranged to form starch, proteins, fats, alcohol, DNA, enzymes, vitamins, and so forth. You, your pet iguana, and all living things are made and kept alive primarily by substances derived from sugar. Man-made materials such as polyester, plastic, and so forth, can also trace their origins back to the lowly glucose molecule.

Fossil fuels also come from this plant-made sugar. This is where it all got started. Figure 8.2 shows just what can be made from sugar.

Here is the important point: all of these various derivatives of glucose carry within them the solar energy that was originally captured by the plant and put into the sugar molecule. All of them, therefore, can have that energy extracted from them, one way or another.

This is the first half of the carbon cycle—how solar energy is captured by plants by adding water to carbon. Now we need to turn our attention to the second half of the carbon cycle, which deals with claiming and using that energy for our lives, for our convenience, and, right now, in ways that threaten us.

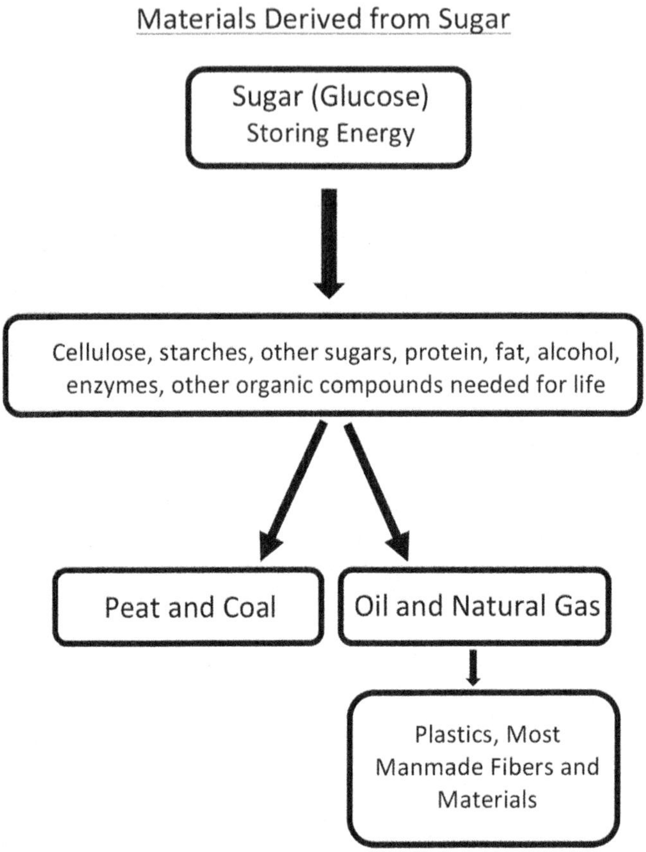

Figure 8.2. Materials Derived from Sugar.

KEY TAKEAWAY POINTS

- Photosynthesis is a chemical process through which energy from the sun is trapped in carbon-based molecules. Carbon dioxide plus water are joined together to form sugar (glucose) while giving off oxygen as waste.
- Sugar and its derivatives are used to build and provide energy for all living creatures.
- The energy in fossil fuels is energy that was captured by plants millions of years ago.

DISCUSSION QUESTIONS

1. "Photosynthesis" as a word is a bit off-putting and "scientific." Can you think of a better, more descriptive term for the process?
2. The statement is made that photosynthesis provides the energy for essentially all life on the planet. Does this seem right to you? Are there forms of life that you think must be exceptions?

Chapter Nine

How Is a Saturn Moon Rocket Similar to a Cow?

Respiration and Combustion

Preservation of our environment is not a liberal or conservative challenge: it's common sense.
—President Ronald Reagan, January 25, 1984 (State of the Union address)

A cow and a rocket have an important similarity. Both run off of solar energy. They both disassemble organic molecules and retrieve the sun energy that was captured through photosynthesis, originally in a sugar molecule. We call this gradual, controlled process "respiration" in biological systems and the faster process "combustion" in nonliving systems. But the overall result is essentially the same.

RESPIRATION

Every sugar molecule, whether made in the most manicured garden, junk-filled empty lot, rain forest, or coral reef, is like a tiny little battery containing the energy of the sun. That energy can be retrieved as necessary to fuel the plant's living processes. It can also be used to fuel rocket ships—but we'll get to that in a minute.

How does the plant retrieve the energy? It pulls in some oxygen from the air, adds it to sugar, and releases the stored energy. Water and carbon dioxide are the waste products for this reaction. If you are paying attention, you will notice that this is the opposite of photosynthesis.

$$C_6H_{12}O_6 + 6\ O_2 \rightarrow 6\ CO_2 + 6\ H_2O + \text{energy}$$

Figure 9.1 is a diagram of this process.

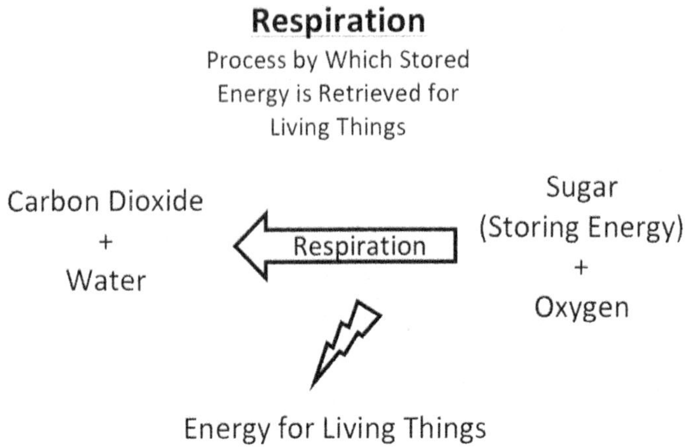

Figure 9.1. Respiration.

What we now have is a cycle—a carbon cycle. It is constant motion, with plants trapping sun energy within the sugar molecule. Figure 9.2 is a diagram of the full cycle.

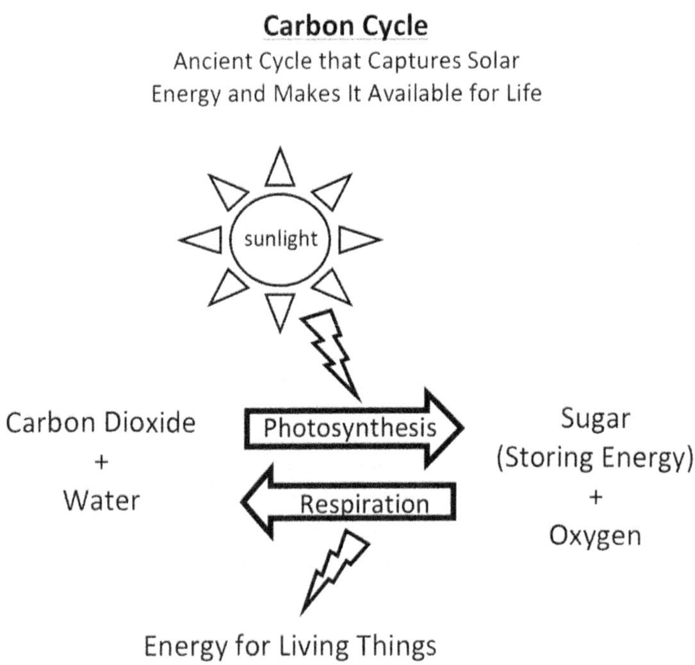

Figure 9.2. Carbon Cycle.

Both plants and animals liberate that energy and use it for living purposes, returning carbon dioxide and water to the atmosphere. The top direction of this cycle (the top arrow) is called "photosynthesis"; the bottom is "respiration."

The word "respiration" is also used to describe the gas exchange in your lungs. Same word but a different meaning. Sometimes the bottom half of this cycle is called "cellular respiration" because it takes place within the cells.

This cycle is how we get our energy from the sun.

Plants are capable of the entire carbon cycle: photosynthesis and respiration. Animals, on the other hand, can only perform respiration and cannot gather energy directly from sunlight. Lying out on the beach may feel warm and produce a sunburn, but that sunlight cannot keep your heart beating or your brain thinking.

You need a go-between. You require a source of preformed sugar, or sugar-derived, energy containing compounds such as proteins and fats. You add oxygen, produce CO_2 and water as waste, and harvest that precious sun energy.

Why Do We Need to Drink Water?

The question comes up—if we give off water as a waste product, why do we need to drink water? In a very real sense your body is a cooperative unit of several trillion individual living things known as cells. Water is a waste product given off by the individual cell. However, you maintain an ocean of sorts to service and surround the cells. The water given off by the cells through respiration is not adequate to provide for this ocean. Hence, drink water.

Much of the sugar is converted into a vast array of other biological compounds used for structure and all the other biological compounds needed for life. All still contain the energy of the sun, and all will eventually give up that energy through respiration, or another process called combustion. (See next section.)

Sometimes you get your energy directly from the plant, like when you eat a nice green salad or rice pilaf. Sometimes there is an intermediary such as a cow, a chicken, or a fish. But originally it all comes from the sun and is collected by plants.

COMBUSTION

The conversion of the plant's energy is not always done gradually for the benefit of the cells of a living organism. Under the right conditions this process can happen rapidly—that is called combustion. Whether a forest fire, or a cozy log burning in the fireplace, you can see and feel that energy being released while quantities of CO_2 and water (which is immediately vaporized) are produced. The net reaction is essentially the same and requires oxygen. The most effective way to put out a fire is by depriving it of oxygen because that ends the combustion process. (Think candle snuffer, fireman's hose, covering a fire with dirt, flame-retardant foam, or a heavy piece of cloth.)

Fossil fuels including oil, coal, gas, and peat can also be combusted and provide energy while using oxygen. They also produce water and CO_2. Anything you find in nature that you can burn for energy can trace its origins back to photosynthesis.

In both cellular respiration and combustion, chemical potential energy is being converted to kinetic energy. Our climate change crisis is a result of the unbalancing of this carbon cycle.

The Rest of the Carbon Cycle

There is a slower part of the carbon cycle that deals with the carbon within the earth and stored in rocks that deteriorate, or build over time. Later, when we discuss other possible sources of CO_2 in the atmosphere, we will refer to carbon from these sources.

KEY TAKEAWAY POINTS

- Sugar and organic compounds derived from sugar can be used to provide energy by reversing the process of photosynthesis. Oxygen is added to the molecule, retrieving the stored energy and producing CO_2 and water as waste products.
- Plants and certain other life forms are capable of both photosynthesis, collecting energy from the sun, and respiration, retrieving that energy.
- Animals cannot perform photosynthesis. They are totally dependent on plants for the energy needed for life, which they garner through respiration.
- Combustion is the rapid conversion of organic energy containing molecules to CO_2 and water with a release of kinetic energy.

DISCUSSION QUESTIONS

1. Looking at the carbon cycle equations—what changes occur to the gases in the atmosphere when a forest is clear-cut or burned down for purposes of human development?
2. It has been stated that all life runs on solar energy. Can you think of exceptions to this? In your life or others', is there an example of life surviving without benefit of energy that originally came from the sun? What would happen if the sun "turned off"?

Chapter Ten

How Is a Saturn Moon Rocket Different from a Cow?

Fossil Fuels

> *The popular notion is that Americans are addicted to fossil fuels, but I find that's not true; most people would be happy to power their lives with anything else.*
>
> —Bill McKibben

Both the cow and rocket run off of solar energy, but in the case of the cow, it is light that shined on the earth and was captured by grass within days, weeks, or months of being consumed. The rocket, on the other hand, is using ancient sunlight that fell on the earth millions of years ago. The energy has been preserved in fossil fuels, which raises the question:—How?

ANCIENT LIGHT

On May 6, 1950, two brothers digging in a peat field near the village of Tollund, Denmark, came upon a body buried about six feet down. Assuming there had been a murder, they notified the authorities. The police recognized, however, that this was not a recent death and contacted the local museum to have it investigated.

The Tollund Man, as he came to be called, had lived during the Roman Iron Age, probably around 400 BCE. He was remarkably preserved, his eyes closed, the one-day stubble on his chin obvious, the leather hat on his head intact.

He is not the only one. There have been several bodies, called "bog" bodies, that have been found extremely well preserved despite interments of thousands of years. Since the culture from which they came cremated their dead, it is speculated that these individuals were sacrifices to the gods, who may have lived in the bogs. The Tollund Man has a braided leather rope around his neck, and forensic scientists agree he died of hanging, likely after eating a ceremonial last meal consisting of an unusual combination of foods.

What has caused these bodies to be so well preserved? As a rule, when a living thing dies, its body is quickly turned into a source of energy for other living things. Decomposition basically means the plant or animal is consumed by animals, fungi, bacteria, and other creatures that gain energy from the dead material through respiration, with CO_2 and water given off in the process. You rarely see dead animals in a forest, or the bodies of birds in the city because nature is quite efficient in recycling these resources.

These "bog" bodies survived because the conditions needed for microbial respiration were not present. The bodies went into an area of "peat" where there was little oxygen available and therefore microbial growth was slow or absent. Decomposition was stopped or slowed dramatically.

These are the same conditions that have brought about fossil fuels.

FROM PLANT TO GAS TANK

The natural life cycle of both plants and animals is a beginning, life, death, and decomposition, which provides building blocks for new life. A cycle.

Plants that died in marshy areas, or in oceans, sometimes were covered with sediment or oxygen-poor water. Because of a lack of oxygen, respiration was not possible, decomposition was slowed down, and the material collected. Covered, and under pressure for vast stretches of time, this ancient plant material became concentrated storage of ancient sunlight energy.

Your car, your home's heater, and much of your electricity run on solar energy—but it is solar energy that shined down on the earth millions, even billions of years ago. Fossil fuels aren't called fossils for nothing.

Fossil fuels can be put into four general categories: peat, coal, oil, and natural gas. Each comes from ancient plant material. Each was produced under different conditions.

PEAT—THE YOUNGSTER OF THE GROUP

The Tollund Man, like eons of plant matter, was buried in a peat field or "bog" where decomposition had been greatly slowed down. When we think of fossil fuels, "peat" does not come immediately to mind. But it is indeed an important fossil fuel both in its own right and as a precursor to another major energy-rich material—coal.

Peat forms when plant material in wet areas is inhibited from decaying because of the marshy conditions. As it accumulates it can create larger wetlands and an increased environment for more peat accumulation. Peat usually builds up slowly, at the rate of about a millimeter per year. Most peat occurs in areas where temperatures are warm enough to encourage vigorous plant life, while being cool enough to allow the collection of oxygen-poor water into permanent wet areas and discourage bacterial growth. Current peat bogs are believed to have been forming for about 360 million years.

It is a wet, spongy material, but when dried, it can be burned and used for fuel, releasing long sequestered CO_2 and ancient solar energy. The brothers that found the Tollund Man were collecting peat to be used for fuel in their home. In several countries, including Ireland, Finland, and Scotland, it is harvested on an industrial scale.

Often it is used as a soil conditioner, known as "peat moss." It is, however, not moss, which is a plant. It is fibrous organic material that can loosen up and regenerate soil for planting.

It can present problems. Peat fields that have dried enough to allow some oxygen into the bog can be a major fire hazard if some of it ignites. Fire from the initial heat can vaporize more water and make more peat available for combustion. The result can be, depending on the size of the peat field, a fire that can go on indefinitely, often underground.

Peat is being replaced by nature at a slow rate. It is being used far beyond its replaceable amounts, and therefore, is a fuel source that is not sustainable.

COAL—CHILD OF PEAT

Over time, peat fields were often covered with sediment from flooding and other geologic activity, further reducing available oxygen and putting the peat under pressure. Millions of years of sediment accumulation, increasing pressure, and heat caused chemical changes that increased the relative amount of carbon, called "carbonization." It became a hard, energy-rich stone that we now call coal. Coal has had much of the hydrogen and oxygen removed from it, so when it is burned in oxygen a greater ratio of CO_2 to water is given off. This means it provides more energy and more CO_2 than a like amount of sugar, or wood, or contemporary plant material.

Coal goes through successive stages, all of which are present today.

- Peat
- The "youngest" coal is called "lignite." This is a brown substance and the lowest "rank" of coal. Today it is burned primarily for producing electricity.
- Subbituminous coal is also used primarily for electricity and in coal-fired plants.
- Bituminous coal is dense, usually black, and used for electricity. This is the most widespread form in the United States. It dates from about three hundred million years ago.
- Anthracite is the highest-ranked coal, is black and heavy, and is used more for heating of homes and buildings. It is found mostly in Pennsylvania; most supplies at this point have been exhausted.

Coal is found on land in areas that once were marshy. It exists in underground formations called "coal seams" or "coal beds." A coal seam can be as thick as ninety feet and as long as 920 miles (1,500 kilometers). This is no small amount of ancient vegetation—it has been collecting for a staggering amount of time.

It is mined with shafts, where men go deep into the earth and drill it out, or if it is less than two hundred feet underground, something called "surface" mining takes place where heavy machinery removes great quantities of dirt to reach it. This is sometimes called "mountaintop removal mining." This takes far less labor than traditional mining. While surface mining is significantly cheaper, the environmental impacts are quite serious. Whole ecosystems are torn apart, destroying habitats, releasing toxic substances that can get into aquifers and the air, and sometimes causing landslides.

OIL AND NATURAL GAS

Where peat and coal occurred primarily on land in wet, marshy areas, oil formed under the oceans. It began as mostly microscopic plants and animals that died and fell to the bottom of the ocean. In very cold, low-oxygen environments they did not decompose and, as with peat, accumulated.

Under great pressure and over time, all that cellulose and organic material evolved into crude oil. The longer this material was under pressure, the lighter the oil, and under the most extreme pressures, the oil changed into natural gas. Oil and natural gas are found under oceans, or where oceans used to be, such as the Middle East, Russia, Texas, and Southern California.

Additional fuels including kerosene, gasoline, and diesel, as well as plastics and numerous man-made materials, come from oil and natural gas; all contain energy.

The fossil fuels—even peat—took a long time to accumulate. We have been releasing the energy and carbon that has been sequestered for millions of years in one or two hundred years. Is this a problem?

KEY TAKEAWAY POINTS

- The energy in fossil fuels is sun energy having come via photosynthesis.
- Fossil fuels were created when the natural process of decomposition was slowed due to a lack of oxygen, and great deposits of plant material accumulated. Over time and under pressure, this became peat, coal, oil, and natural gas.
- These materials are a vast reservoir of energy that took millions of years to form and accumulate.

DISCUSSION QUESTIONS

1. What is it about fossil fuels that has caused them to be the energy choice for modern civilization?
2. Are these materials—peat, coal, oil, and natural gas—still being created in nature?

Part III

Global Warming and Climate Change—Why Should We Care?

Chapter Eleven

Giver of Life

The Greenhouse Effect

I love the feeling of fresh air on my face and wind blowing through my hair.
—Evel Knievel

It's a trite question: "Is a glass partially filled with water half-empty, or half-full?" Presumably, how we answer gives insights into our character. In fact, the correct answer, regardless of one's personality, is "The glass is full."

Air is easy to take for granted, common, constant, and, of course, usually invisible. We can lose sight of the importance of the gaseous ocean in which we live. A windy day, a good storm, or holding one's breath for thirty seconds provide reminders. But our atmosphere is more than just an oxygen source or provider of water and weather.

There are people who have lived in places without an atmosphere. Perhaps, to get a better understanding of the benefits of gaseous living, we should see how they accomplished it.

DRESSED FOR SUCCESS

The outfits our moon invaders wore on the job may not have been haute couture, but they were remarkable pieces of technology, designed for the specific rigors of a world without air. Made of twenty-one layers of specially designed materials, they included a ninety-pound backpack. Overall the things weighed 180 pounds on Earth, but mercifully, the reduced gravity on the moon brought that down to a manageable thirty pounds or so.

What do these space suits need to do? First, just like our natural gaseous environment, they provide the necessary oxygen as well as a way to collect the carbon dioxide that the astronaut exhales. Second, the suit must be tough enough to withstand increased amounts of gamma radiation and tiny bits of space debris, little of which makes it through our atmosphere.

The suits need to provide pressure. It doesn't feel like it, but human bodies are designed to live under a great deal of pressure, which is essentially the weight of miles of atmosphere pressing down on us. We are in equilibrium; that is, our bodies exert the same amount of pressure outward. Take away the outside pressure and . . . it would not be pretty. So, a space suit, often called a pressure suit, is designed to exert a pressure on the astronaut's body the same as the atmospheric pressure on Earth.

Finally, and most important to our discussion of climate change, is the all-important temperature management. Humans, like most living things, only do well within a fairly narrow range of temperatures. The moon, as well as space in general, offers serious extremes. On the sunny side of the moon temperatures can reach around 250° Fahrenheit. Hot enough to boil water or any handy astronaut. With no atmosphere to inhibit it, that energy is reflected rapidly back into space; so, were the astronauts to step over onto the dark side of the moon, the temperature would drop precipitously to something like −250°F. Very, very chilly.

Our moon walkers stayed on the sunny side, so they carried air conditioners on their backs. That equipment was designed to maintain a relatively comfortable living temperature.

We don't have those kinds of temperature changes on the earth between night and day—why not? We are heated from exactly the same sun, but one of the most important functions of our atmosphere is a moderation of the temperature around the planet. It gets cooler at night, and warmer by day, but not nearly the extremes we would experience without an atmosphere. Without this moderating function the extreme temperature fluctuations would preclude essentially any kind of life.

More than just preventing extreme temperatures, however, our atmosphere, like the thermostat on an astronaut's space suit, keeps the world at a generally comfortable life-sustaining temperature. Like a blanket, it keeps us warm. How?

THE GREENHOUSE EFFECT

In an effort to understand our relatively comfortable temperatures, John Tyndall, in the 1860s, did a pioneering study in which he bombarded various gases with infrared radiation to see how much heat each gas would retain. Our air is mainly made up of nitrogen and oxygen, but these two gases do not hold any appreciable heat. If we only had these two gases in our atmosphere, our earth would have an average temperature of about −15°C or 0°F. Very icy and precluding most life as we know it.

Tyndall found that both water and carbon dioxide held heat. He identified them as the likely candidates to be keeping our air and planet at a comfortable temperature. His work has been verified with more precision and today we know several other gases retain heat in the atmosphere: methane, nitrous oxide, ozone, and chlorofluorocarbons (CFCs). But water and carbon dioxide are still the most significant holders of heat.

Energy comes from the sun as electromagnetic radiation, mostly in the visible range. Some of it is immediately reflected back into space as infrared radiation. Some of it bounces around, does work, smashes tomatoes, grows cows, that sort of thing, but eventually it, too, dissipates into the atmosphere as infrared radiation. The "greenhouse effect" is a result of some of this radiation being held by small amounts of these gases. Collectively, these gases are referred to as greenhouse gases, or GHGs for short.

With the exception of water (which is highly variable), these gases make up less than 1 percent of our atmosphere. Though their amounts be small—their effect is mighty; they have made our planet livable. Of these, the most important is CO_2, not because it holds the most heat—that would be methane—but because it is in much larger amounts than the others and, once in the atmosphere, it can stay for many centuries.

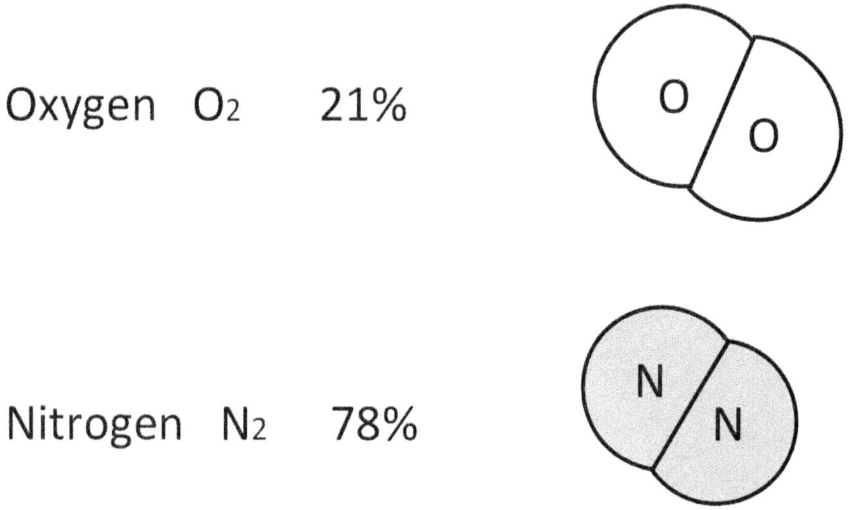

Figure 11.1. Major Gases in the Atmosphere.

For the last eight hundred thousand years, for which we have data, the GHG levels have been relatively stable, keeping our world within a narrow range of temperatures. What could possibly go wrong?

KEY TAKEAWAY POINTS

- We live in a gaseous ocean that is essential to life in several crucial ways.
- In addition to providing gases necessary for photosynthesis and respiration, our atmosphere moderates the temperature of the planet and maintains it at a warm enough temperature for life to thrive.
- Certain gases, though in small quantities, hold on to significant heat, which accounts for the greenhouse effect that protects us from the freezing temperatures of space.
- These greenhouse gases (or GHGs) include methane, nitrous oxide, ozone, chlorofluorocarbons, carbon dioxide, and water.

DISCUSSION QUESTIONS

1. Describe what you think life would be like in an atmosphere half as dense as ours. Half as much nitrogen, oxygen, water, other gases. (Hint: read some accounts about climbing Mount Everest.)
2. What if oxygen, 20 percent of our gaseous ocean, was a GHG. How would this change things on Earth? What if nitrogen, about 79 percent, was a GHG?

Chapter Twelve

What Could Possibly Go Wrong?

There is no energy crisis, only a crisis of ignorance.

—R. Buckminster Fuller

In the 1960s, Andy Warhol became one of the most famous artists on the planet with his paintings of consumer items. Rows upon rows of Campbell soups, Brillo Pads, bottles of Coca-Cola. Even portraits were in multiples. Why have a single image of Marilyn Monroe when you can do twenty?

Some found him to be trend setting, "brilliant." Others wrote him off as a joke or insulting to art. Regardless, there was no denying that he mirrored the times. His art reflected a world obsessed with consumption and mass production. Just as his work was dramatically different from what was considered acceptable art by the "French Salon" a hundred years earlier, the world he portrayed could scarcely be imagined by individuals from 1860. His art was symbolic of a rapid upending of global culture.

The nineteenth century ushered in dramatic changes to human lives, changes that couldn't be anticipated at the time. In the Industrial Revolution we learned how to mass-produce more and more material goods with unheard-of speed. Our transportation went from the horse and buggy to smoke-belching trains, eight-cylinder road-hugging machines, and airbuses crisscrossing the skies thousands of times a day. Cities, once sedate and quiet, if a bit smelly and muddy, suddenly lit up in blazing light, and became noisy with a cacophony of vehicle sounds, motors, and loudspeakers. And the gadgets, so many gadgets.

We now have many more people, who are buying much more stuff, traveling more, and using increasingly large amounts of electricity—a perfect storm that has escalated our energy usage exponentially.

What cheap, abundant source of potential energy have we turned to to power all this light, action, and abundance? We have dug deep into the earth and extracted great quantities of long sequestered ancient plant material in the form of coal, oil, and natural gas. Bit by bit, we have designed a world dependent on ever-increasing quantities of fossil fuels to provide comfort, convenience, lots of stuff, and a shadow over our future.

THE FIRST HINTS

There were a few voices from early in the Industrial Revolution that foresaw the possibility that human activity could affect the natural world.

George Perkins Marsh, in 1847, was the first person to suggest that human activities could affect the earth's "balance of nature." In 1894, Svante Arrhenius, a Swedish chemist, suggested that the increased emission of carbon dioxide from the Industrial Revolution could lead to global warming. He predicted that in four thousand years the CO_2 in the atmosphere could double, raising temperatures 5°F. His temperature estimates weren't too bad, but his timing was way off; he vastly underestimated both the degree of industrialization and the population explosion.

In the late 1950s, at the Mauna Loa Observatory in Hawaii, Charles David Keeling began measuring the levels of carbon dioxide in the atmosphere and resulting changes in climate. Those measurements and his diligence gave rise to the modern realization of man-made global warming,

In the 1970s, Dr. James Hansen was a mild-mannered NASA scientist studying the curious situation of our "sister" planet, Venus. Despite the fact that it is closer to the sun, less sunlight reaches the surface of the planet because it is surrounded by a heavy cloud cover. Nonetheless, its surface temperatures are around 874°F, or 462°C. Lead would melt. Why? It turns out that the atmosphere on Venus is largely CO_2. Venus is a cautionary tale of the power of a runaway greenhouse gas.

Dr. Hansen, in the early 1980s, turned his eyes from Venus to the home planet and the effects of the growing CO_2 in the atmosphere of our own planet. He began to speak out about this increasing CO_2 and its effects on the climate and natural life cycles.

It has taken a long time for scientists, who are natural-born skeptics, to come to consensus about the effects of rising CO_2 in the atmosphere. But as methods of measurement have improved and the evidence has accumulated, the scientific community has come to the unquestionable conclusion that the earth is warming, global climate is being disrupted, and it is due to the accumulation of anthropogenic GHGs, primarily carbon dioxide.

TOO MUCH OF A GOOD THING IS NOT A GOOD THING

Today we are releasing millions of years of sequestered CO_2 rapidly into our air. The carbon cycle, in balance for so long, has been weighted on one side with more CO_2 entering the atmosphere than can be removed by the natural photosynthetic processes. It is too fast for the natural evolutionary process of geology, or life, to adapt.

There is an added problem. Common sense suggests that we ought to be looking for ways to increase plant life, to increase photosynthesis and begin reabsorbing some of that CO_2. Unfortunately, just the opposite is happening. Rain forests, the so-called lungs of the planet, are being burned and destroyed to make way for development, grazing of animals, and palm oil production. We need healthy ocean vegetation, but our oceans are one of the first parts of the planet to suffer from the effects of too much CO_2 as well as pollution and overfishing. In many parts of the ocean, this source of photosynthesis is being reduced.

Closer to home, our modern way of life and industrial production are encroaching on plant life. A natural field or even a weed-covered plot of land is pulling in CO_2 from the air, and releasing oxygen. A tidy-looking parking lot or modern strip mall is not.

CO_2 is being produced in larger amounts and absorbed in smaller amounts. The result has been a gradual buildup of CO_2 and change of chemistry in the air. Figure 12.1 is a diagram that shows the change in the carbon cycle with fossil fuels added in.

Several methods have been used to determine CO_2 concentration in the atmosphere over the last eight hundred thousand years. Ice core data has been some of the most definitive. Natural factors have kept the range within about 180 to 300 parts per million (ppm). Warmer periods have correlated with higher CO_2, close to 300 ppm, concentrations. For the last forty thousand years the level has been close to 275 ppm, which is what it was at the beginning of the Industrial Revolution. In 2012, the CO_2 level hit 400 ppm for the first time, higher than has been seen since long before humans walked the earth, and it continues to climb.

The cause of global warming can be summarized in three points:

1. Carbon dioxide holds warmth in the atmosphere.
2. We are adding large amounts of carbon dioxide to the atmosphere, beyond what the natural cycles can remove.
3. As a result, the atmosphere and the planet are warming.

These points are not in dispute. This is basic chemistry and physics. The natural greenhouse effect, which has been in balance for thousands of years, protected us, and made our planet habitable, has been put on steroids.

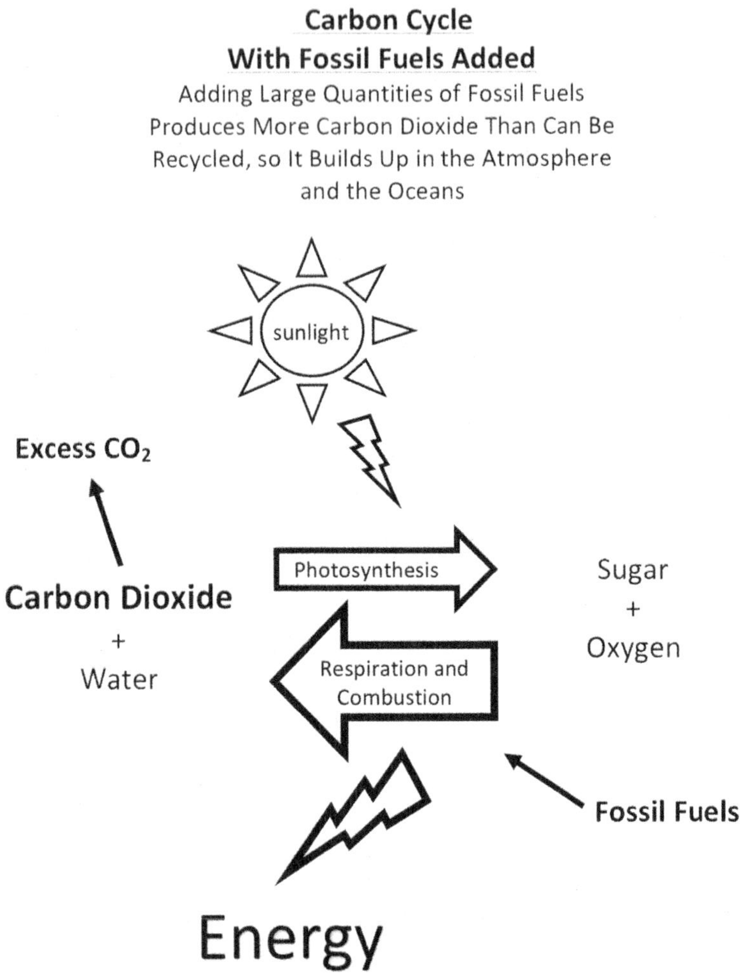

Figure 12.1. Carbon Cycle with Fossil Fuels.

OTHER GREENHOUSE GASES

While carbon dioxide is the most significant of the greenhouse gases, it is not the only one that is being affected by human activities.

Methane, the primary component of "natural gas," is second in importance. This gas can hold up to twenty-five times as much heat as CO_2 but is found in considerably smaller amounts and stays in the air for less time. Nonetheless, its contribution is significant, probably about one-third that of CO_2.

Natural sources of methane include peat bogs, swamps, rice paddies, and other wet places with relatively low oxygen. Other sources are industry, wastewater, and cattle that produce it when chewing their cud. Landfills also produce methane. This gas still contains energy; there are a number of efforts to capture the gas from landfills both to keep it out of the atmosphere and to burn it for heating or electricity production.

Ozone is a gas of different personalities. The oxygen we breathe is O_2, with two oxygen atoms sharing space. Adding a third makes it ozone, and, unlike its doublet cousin, this is harmful to breathe.

Everyone has heard of the beloved "ozone layer," which is a layer of this gas in the stratosphere that shields the earth's surface from harmful levels of ultraviolet light. The damage to the ozone layer has often been conflated with global warming. It is important to note that these are not the same thing and have only a passing relationship. Ozone is obviously of concern regarding protection from UV light, but it is only a bit player in the global warming drama.

For clarification, it is worth distinguishing between these two phenomena. In the 1980s it became apparent that chlorofluorocarbons, man-made chemicals used in spray cans and refrigeration since 1920, were eating away at the protective ozone layer in the stratosphere. In 1987 the Montreal Protocol, proposed by the UN, was ratified with global support, to replace these chemicals. Since then the ozone layer has stabilized and is expected to rebuild over the next several decades. It is an example of international cooperation that benefits all.

Prized as ozone is at the upper reaches of the atmosphere, at lower levels, it is a strong GHG and pollutant, often associated with stagnant air and smog. Short lived, and not well distributed in the atmosphere, ozone concentrations are difficult to measure with accuracy. Lower levels are believed to have increased perhaps 30 percent since the Industrial Revolution.

Chlorofluorocarbons are also powerful long-lived greenhouse gases; levels were increasing rapidly in our atmosphere until we began to phase them out after the Montreal Protocol. Since then they have begun to level out. Despite their large heat-holding capacities, they still make up a very small part of our atmosphere.

Nitrous oxide is a waste product of industry and agriculture, particularly chemical fertilizers. Also capable of holding significant heat, it lasts about a century in the atmosphere, but is still in small amounts. Those amounts are increasing by about 1 percent a year. Changing agricultural practices can reduce the release of this gas.

Water is also a GHG. The amount in the atmosphere varies a great deal due to the dynamic nature of the water cycle. Human activities have not been directly responsible for added water vapor. But there is an ironic twist here. As the temperature of the earth and air rise, more water evaporates, increasing its levels, which, of course, helps increase temperatures. This is one of the more concerning parts of the global warming situation.

CO_2 remains the most serious culprit. But it is important to recognize the contributions of these other gases, which are very real.

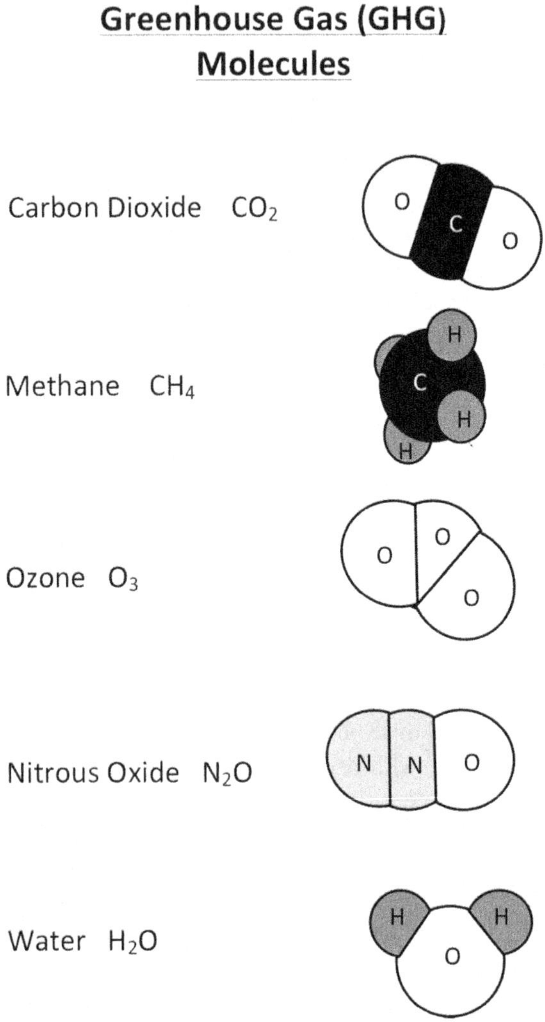

Figure 12.2. Greenhouse Gases.

KEY TAKEAWAY POINTS

- The Industrial Revolution ushered in a time of unprecedented consumption, travel, convenience, and the need for vastly increased quantities of energy.

- The reliance on fossil fuels, releasing millions of years of sequestered carbon, has resulted in unprecedented increases of carbon dioxide and other greenhouse gases in the atmosphere.
- The natural carbon cycle is unable to recycle all these gases; as a result they are building up and resulting in an enhanced greenhouse effect that is warming the planet.

DISCUSSION QUESTIONS

1. In a given day you use a lot of things, eat various foods. List five and explain the role of fossil fuels in the production of each. Can you list five items you regularly use that needed no fossil fuels in their manufacture or growth?
2. How does transportation in the eighteenth century compare with transportation today? How is it similar? How is it different? How has the amount of transportation a person uses changed?

Chapter Thirteen

So What?

Consequences of a Warming Planet

> *By burning fossil fuels, we are dumping 30 billion tons of CO_2 into the atmosphere every year, which has a profound effect on the climate. So, like it or not, we're already messing with a system we don't understand.*
>
> —Jeff Goodell

There is no disagreement within the community of atmospheric scientists and geologists with the basic realities of global warming:

1. Carbon dioxide (as well as other GHGs) in the atmosphere holds heat.
2. We are adding more carbon dioxide than the natural processes can remove.
3. This additional CO_2 in the atmosphere is warming the planet.

Disagreements come from how fast we will be seeing the effects, how severe they will be, and the best ways to address them. There is a range of possibilities, and, while we have some very good models and predictions, in a system as large as the earth, there remains a fair amount of uncertainty. The Intergovernmental Panel on Climate Change (IPCC) was created in 1988 with the express purpose of informing governments of the latest scientific information on our changing climate. It represents the world's best atmospheric and climate scientists and the most current data, but always provides a range of projections that depend both on variances and unknowns. Earth cycles and human activities are complicated, and the panel tries to account for different levels of response by the global community.

Nonetheless, certain changes are occurring now and more can be projected for the future. Many details are not precise, but the general effects can be said with certainty.

The IPCC

The most comprehensive resource on global climate change is the IPCC (Intergovernmental Panel on Climate Change). The IPCC was set up in 1988 by the World Meteorological Organization (WMO) and the United Nations Environment Program (UNEP) to provide policy makers with the scientific basis of climate change, its impacts, future risks, and options for adaptation and mitigation. It involves the work of thousands of scientists from all over the world. Membership in the IPCC is open to all member countries of the United Nations (UN) and WMO. Currently 195 countries are members.

It publishes regular assessments with updated information and is currently working on the 6th report. The 5th assessment report was published in 2013, and is available along with other information about the IPCC and its activities here: http://www.ipcc.ch/index.htm.

The report is long and technical, a bit of a slog to get through. A 2015 book by Michael E. Mann and Lee R. Kump, *Dire Predictions: Understanding Global Warming; The Illustrated Guide to the Findings of the IPCC*, summarizes the findings and is an easier read.

SO FAR

Prior to 1850, the earth's average temperature had been stable for one to two thousand years. Since then, it has been moving up at an unprecedented rate. The overall increase in temperature since 1880 is about .85°C or 1.5°F. Different parts of the earth have warmed more than others; some have actually cooled. That is an average of land, atmosphere, and ocean changes and is based on multiple independently produced databases.

The rate of warming is increasing, and the twenty-first century is projected to see a rise of from 1.7°C to 4.8°C (3.1°F to 8.6°F) (relative to preindustrial temperatures) depending on what efforts are made to reduce global emissions of CO_2. These are temperatures not experienced since the mid-Pliocene, around three million years ago. Even a rise above the preindustrial level would be outside the range of temperatures experienced by human civilization. Right now, limiting the temperature rise to 2°C is the goal of the IPCC.

We are used to daily fluctuations greater than this and it is hard to see this as the serious problem it is. Changing the average temperature of the entire planet—land, air, and sea—by these amounts represents a tremendous gain in heat and carries with it serious consequences. A major part of the problem is the *unevenness* of the temperature rise, which is a major driver for changes in climate.

These relatively small increases in global temperature are already producing effects. For most of us it may feel a bit warmer in summer, a slightly shorter winter, or perhaps we notice a greater range of temperatures. There have been extended droughts, unusual flooding, wildfires, and more extreme hurricanes and other weather incidents, but if you don't live in one of the affected areas, it feels manageable.

For the time being we can write off the consequences as pretty much happening to "others." In other words, denial isn't too difficult. Global warming, however, will ultimately affect everyone.

At some point, changes are expected to generate feedback loops, which means that the increases in temperature result in unabated future increase regardless of human activities.

COMING ATTRACTIONS

A brief summary of further consequences is offered here. Additional in-depth explanations are available online, in books listed in the bibliography, and, of course, the IPCC reports.

Virtually every one of the earth's interrelated cyclic systems is being affected by the imbalance of the carbon cycle and the resulting warming of the planet. Let's begin with that all-time favorite conversation starter—the weather.

THE WEATHER AND THE CLIMATE

Occasionally the question comes up: "Is it global warming, or is it climate change?" Climate change is one of the symptoms of global warming and better explains what people are experiencing. Even though the earth is warming, in many areas, the climate is actually bringing increasing cold weather. In others weather is becoming less reliable. The term "climate disruption" is often used and is probably the best description of what is happening. These terms are used interchangeably.

The earth's water cycle is changing in a number of ways. Warmer temperatures cause more water to evaporate into the atmosphere. Certain areas lose more water and can experience drought. But more water in the atmosphere brings about bigger and wetter storms in other areas. The uneven distribution of warmer air has altered some of the normal pathways of storms and weather.

Areas that depend on rainwater or snowmelt for agriculture may find themselves with less. Other areas are receiving more snow, rain, and colder temperatures. This often causes people experiencing cold to question global warming.

One of the more humorous, if ill-advised, misunderstandings of this fact was when Senator James Inhofe, from Oklahoma, tossed a snowball on the Senate floor in the winter of 2014. He insisted that the unusual snowfall in Washington, D.C., that year was proof that global warming could not be happening.

In fact, if he was demonstrating anything, it was just the opposite. The unusual cold that city was experiencing was one of the projected weather disruptions consistent with climate change. It is worth noting that at that very moment, Florida was experiencing record winter highs, California had its smallest snowfall in decades, and 2014 was the warmest year to date for the globe. That record was broken in 2015 and again in 2016.

In the past several decades we have been experiencing a significant increase in extreme storms, flooding, and heat waves. Scientists, cagey as they are (see chapter 15, "The Trouble with Scientists"), are unwilling to state that a single event is, in fact, a result of global warming. As is often pointed out, there have been extreme weather events throughout history. What they will say is that the increasing frequency of these events is consistent with what is expected with global warming.

As the planet warms more, putting more water into the atmosphere and changing the atmospheric and oceanic currents, scientists predict with confidence that extreme weather events of all kinds will become considerably more common.

DRIP, DRIP, DRIP: ICE IS MELTING, OCEANS ARE RISING

Santa Claus made an unfortunate choice when he decided to set up camp at the North Pole. In terms of travel needs it made sense; more people, including starry-eyed small children, live in the Northern Hemisphere than the Southern. But unlike the South Pole, the north has no land mass under it. As increasing temperatures melt the ice, his frozen real estate is disappearing and all those workshops and reindeer abodes are under threat.

Interestingly, the diminishing North Pole ice mass does not make a great deal of difference to the amount of water in the ocean. Since ice—solid water—is unusual among substances and is lighter than liquid water, when it melts it does not add to the volume of the water. Try it yourself. Put ice cubes in a glass of water, mark the level of the water, and let it sit around for a while. The level of the water will change very little with the melting of the ice.

We cannot blame rising sea levels on a melting North Pole. This, however, is of small comfort to assorted aquatic and land animals that live there, including polar bears and of course, Santa. It is an ecosystem that is being reduced, compromised, and eliminated. We are only beginning to understand the consequences of a melting and warming Arctic on the health of the oceans.

The earth stores vast, almost incalculable, amounts of frozen water on land. The South Pole, Greenland, and great glaciers and expanses of ice in the northern parts of Russia, Canada, not to mention the snow-covered mountains in both the Northern and Southern Hemispheres, have been holding staggering quantities of water since long before humans walked the earth. As the earth warms, that water is being liberated to the sea.

There is a second factor increasing the volume of the oceans. Warmer water expands slightly. In small amounts of water, say, your neighbor's swimming pool, this is insignificant. However, in a body of water the size of the oceans, it makes a real difference. Warming adds to the volume of the seas and the rising ocean levels.

So far the effects for most of us are barely noticeable. Currently, sea levels are rising about 3.1 mm per year—that's less than a quarter of an inch. Since 1901, sea level has increased roughly eight inches. Hardly seems worth a bout of hysteria.

That said, there are a number of low-lying islands that have already been impacted. Certainly coastal cities around the world have noticed more common high-tide flooding. Municipalities, including Venice, Miami, and Boston, have already begun planning and implementing projects, usually costing billions, to hold back the sea as levels rise.

It is important to remember that the oceans have a lot of "lag time." It takes longer for water to warm up than the atmosphere and therefore longer to see consequences. Also, the rate of CO_2 pollution is increasing and, consequently, the rate of sea level rise is increasing. A number of very good models come to different conclusions.

One group of sixteen of the world's leading climate scientists led by James Hansen published a paper on July 23, 2015,[1] concluding that the rise could be an additional ten feet within the next fifty years. Other studies have suggested anywhere from two feet to twenty feet by the end of the century. Some have suggested a sea level rise as much as eighty feet. The IPCC is usually more conservative and sees a range of around three feet by the end of the century.

What is acknowledged by all models is that sea level is rising, that it is doing so at an increasing rate, and that it will continue to rise over the next several centuries.

An increase of ten feet would dramatically impact low-lying coastal plains such as a lot of Texas, North Carolina, Louisiana, Florida, and cities including San Francisco, New York, Los Angeles, Amsterdam, Mumbai; just about any city on a coast or river could be dealing with devastation. Two-thirds of the world's major metropolitan areas are located in places making them vulnerable. Low-lying areas miles from the coast could also be dealing with routine flooding. Even a three-foot rise would cause significant destruction in coastal cities, low-lying areas, and island nations. Costs worldwide could be in the trillions.

As with so many disasters, parts of the world with the fewest economic resources, such as Bangladesh, will witness some of the most serious effects of expanding oceans.

A good summary of the melting ice problem is *Chasing Ice*, an award-winning documentary on climate change by National Geographic.

THE HEALTH OF THE SEA—THE LIFEBLOOD OF THE PLANET

What's not to love about the oceans? Great sunsets, calming waves, and a remarkable variety of seafood. However, they are far more important to human survival than a place to float our cruise ships, thrill surfers, and provide good shrimp cocktail.

To be sure, much of the world's food comes from the ocean, roughly 140 tons per year. So does a lot of the world's oxygen, approximately 70 percent of it, through photosynthesis by aquatic plant life, mainly phytoplankton. The oceans serve as a moderator of temperature for the planet. Liquid water can hold more excess heat than the air, so throughout time, the oceans, like the greenhouse effect of the atmosphere, have helped maintain a livable temperature on the planet and kept it relatively constant.

Our oceans, already under stress from the dumping of human-made waste, everything from the organic leftovers of feedlots, to chemical fertilizers, plastic bags, and nuclear waste, are now dealing with a buildup of CO_2. As carbon dioxide builds in the atmosphere, much of it is absorbed by the oceans. They are currently absorbing about twenty-two tons of CO_2 a day.

Initially this might seem positive—less carbon dioxide in the atmosphere means less warming. And it is true, this is a moderating function of our ocean. However, the quantities of carbon dioxide released into the atmosphere are overwhelming the photosynthetic capacity of the oceans. Excess CO_2 is converted into carbonic acid. It is "acidifying" the sea.

The oceans have maintained a stable acidity level for tens of millions of years within which the rich and varied plant and animal life has evolved and survived. This ancient balance is being gradually undone. The acidity is expected to drop to deeper levels and increase as the ocean continues to absorb excess carbon dioxide.

The consequences of this heretofore unheard-of phenomenon have only recently begun to be understood. We recognize that carbonic acid inhibits shell growth in marine animals, and it is suspected to be a cause of reproductive disorders in some fish.

Warmer, more acidic water has the most effect on the tiniest of animals, the photosynthesizing phytoplankton. Not only do these tiny animals produce most of the oxygen we breathe, they are the basis of the food chain upon which many aquatic species are dependent.

Further effects on marine ecosystems need more study, but all signs indicate that the warming and acidification could be highly destructive, and is happening too fast for organisms to adapt.

Coral Reefs

For some, their knowledge of coral reefs may be limited to the movie *Finding Nemo*, which did accurately portray the reefs as beautiful structures awash with life. They are a combination of animals and plants, and they provide homes to a vast diversity of creatures, including clown fish.

They are huge and protect many shorelines from the ravages of ocean storms, while absorbing a great deal of carbon dioxide and producing oxygen. Having been built over thousands of years, they are now dying. The Global Coral Reef Monitoring Network (GCRMN)[2] reports that currently 27 percent of all coral reefs worldwide have been lost, and without significant intervention to reverse climate change as well as other damaging human activities on the reefs, the outlook for the next thirty years is considered poor. There are some predictions that as early as 2050 all but 30 percent could be gone.

A number of additional human factors are stressing the reefs, including dredging, overfishing, and pollution. But recently it has become clear that the increase of carbonic acid in the water is limiting the ability of corals to form skeletons. Warmer temperatures weaken the corals and encourage the growth of bacteria and other pathogens. It also is causing "bleaching" of the coral, which is the loss of photosynthesizing organisms, usually the first step to death of the reef and the entire ecosystem it supports.

Sea level rise can reduce the amount of sunlight reaching the corals, therefore further reducing their ability to photosynthesize and making formation of the coral structure difficult. The dying of these rich aquatic ecosystems reflects the health problems of the oceans. As the oceans go, so does life on this planet, if not immediately, certainly in the future.

We rarely see below the surface of the sea. But that surface hides a staggering quantity and variety of life that is essential for terrestrial life.

Today a great deal is going on below that surface that does not bode well for life on the planet.

LIVING COMMUNITIES, BIODIVERSITY

Everyone has heard of the term "ecosystem," but not everyone has a good idea of what it is. Basically it refers to a group of organisms that interact with each other and their environment, for the benefit of all, a "living community." In the largest sense we can consider the earth an ecosystem. Within its great cycles, including the cycles of life and death, predator and prey interact and maintain stability in populations and life as a whole.

They can also be tiny—you can buy one, $89.95 on Amazon. It is an ecosystem completely enclosed in glass, containing plants, aquatic animals, and microorganisms that, with a source of light energy, will feed each other and maintain homeostasis through life and death cycles.

The Great Barrier Reef is considered an ecosystem, as is a tropical rain forest, the Arctic tundra, great swaths of desert, or mud pots in Yellowstone National Park. All require an input of energy, usually the sun, a source of water, and so forth. When changes occur to a member of the ecosystem or to its environment, change is required of all other parts. If change is not possible, the entire system can be at risk.

There is no question that the general warming of the planet, changing the weather patterns, the ocean currents and behavior, and causing wildlife to migrate or die out, will have an effect on countless global ecosystems. It is difficult to specifically delineate the changes in the future. But already there is evidence of species die-off and migration.

As the temperature rises, the timing of life cycles changes and certain animals and plants are affected more than others, upsetting the balance. Agriculture will be affected, not just by more droughts and flooding but by changing weather and temperature conditions that may favor some types of vegetation, but in general is expected to be destructive to more.

Virtually all systems are vulnerable.

HEALTH AND DISEASE

One topic often overlooked is that of changes in diseases and public health. In 2009 in the medical journal the *Lancet*, a team of British scientists reported: "In terms of our well-being, in terms of our survival over the next 100 years, [climate change] is absolutely the top public health issue that we should be talking about."[3] They continue to update their global health consequences on the referenced website.

Health-related concerns so far delineated include, but are not limited to, the following:

- As the earth warms, diseases confined to the tropics are expected to migrate north and south. Insect vectors that carry malaria and other pathogens will find a welcome environment on larger parts of the globe.
- Increasing extreme weather, such as hurricanes and drought, bring with them casualties and suffering. Most outbreaks of waterborne diseases occur after extreme weather events including storms and flooding.
- Heat waves, sometime called the "silent killer" because such events don't get all the attention of pathogen outbreaks, can be expected to increase in frequency, duration, and magnitude. The 2003 heat wave in Europe killed an estimated thirty-five thousand people, breaking all records for heat-related deaths.
- Changing weather patterns will disrupt agriculture and can bring food shortages and starvation. Plant pathogens will change and likely increase in range due to warmer, more humid conditions.
- Poor air quality, a result of warmer air, and man-made pollution will bring about more ground-level ozone—smog. Interestingly, as CO_2 concentrations increase, certain plants, including ragweed, produce disproportionately more pollen. Allergy-related disease, already on the rise, is expected to increase.
- Young children and the elderly are generally the most vulnerable to these changing disease patterns. Likewise, people living in poorer nations, without air-conditioning or efficient food distribution, will pay the high price for the fossil fuel indulgence of the richer countries. But it will ultimately affect us all, rich and poor, old and young.

CONSEQUENCES OF THE CONSEQUENCES—TIPPING POINTS

And then there are the feedback loops, where the heating takes on a life of its own without help from humanity. When this kicks in—and scientists don't know when—even the most drastic energy austerity will not be enough to stop continued heating of the earth. Global warming will be off the reservation, feeding itself. Some examples of these feedback effects include the following:

- Snow and ice is white, which reflects sunlight back into space. As the ice of the North Pole is replaced with dark ocean water, more heat is absorbed by the ocean. Likewise, the reduction of glaciers, snow on mountains, and so forth, means less radiation is reflected and heating is increased.

- As the atmosphere warms, more water is evaporated and held in the air—humidity. While this is one of the factors driving more extreme storms, water vapor is also a greenhouse gas, thereby trapping still more heat in the atmosphere.
- The heating of the oceans, as well as other kinds of pollution, is reducing the abilities of the oceans to photosynthesize. This leads to more CO_2 in the atmosphere and oceans, and more warming of the atmosphere and oceans.
- There is something called Arctic methane emissions. As permafrost (ground that heretofore has been frozen) melts, and glaciers in the Arctic ice recede, methane and additional CO_2 are released into the atmosphere. Part of this is just stored gases; part of it is a result of rotting biomass that was previously frozen. As the earth warms, more of these gases are released, which add to warming the earth.

It is this kind of runaway greenhouse effect that is the scariest, threatens the livability of the planet in the future, and keeps climate scientists up at night.

ECONOMICS

Extreme hurricanes are expensive. Droughts, flooding, and wildfires are expensive. Additional costs of public health and health care will be expensive. Protecting major cities from flooding will be ridiculously expensive, and moving them is off-the-charts expensive. Retooling agriculture and our food distribution system will be expensive.

Global warming is going to be expensive. Even if we stopped emitting fossil fuels tomorrow, because of the lag time in the effects, we can expect considerable expenses during the next century. The sooner we address it, however, the more we can mitigate the costs.

One of the arguments against taking action is that it is just "too expensive" to move to sustainable energy. The fallacy of that argument will be addressed in chapter 26. But regardless of what it costs, it would be minuscule compared to the costs of inaction.

WAR AND REFUGEES

Much as we like to think otherwise, when resources are scarce humans forget the lessons they learned in kindergarten. They no longer play nice with others, and they don't like to share. As our capacity to destroy and kill has outstripped our abilities to cooperate, violence may be the greatest threat to our species that continued CO_2 emissions presents.

In a report to the Senate Appropriations Committee dated July 28, 2015, the Department of Defense stated that "global climate change will aggravate problems such as poverty, social tensions, environmental degradation, ineffectual leadership and weak political institutions that threaten stability in a number of countries."[4] It has labeled climate change as a "catalyst for conflict."

Serious as our refugee problems seem to be in 2017, they no doubt pale compared to the numbers of displaced and desperate people that climate change will create.

Paradoxically, the kind of global cooperation needed to address the problem may usher in a new understanding of our interdependence on this planet and help us develop ways to resolve international disputes other than war and violence.

WHO WILL BEAR THE BRUNT OF INACTION?

Future generations will deal with the heavier consequences of today's inaction. One is tempted to say, "Let them deal with it when it gets here." That is a serious misunderstanding of the situation; the earth is huge and does not turn on a dime, and there will be a point where turning is not even possible. By the time children born today reach their midthirties, it will likely be too late for human endeavors to slow this course.

Not doing anything today is like today's adults having a huge party on the top deck of the *Titanic* while our kids play below. Inadvertently, we have set a course toward an iceberg, but there are plans to helicopter us off, so we party on. When we see that ice looming and realize it is too late to change direction, we load onto our transport, Saigon-style, and wave good-bye to the children who are left with no options.

Not that any of us is getting out of here on a helicopter, but right now we are in a place to ensure a future for our children, or to seal their fate.

At a talk given to the French National Assembly on May 18, 2010, Dr. James Hansen said the following:

> It is my job, as a father and grandfather, concerned about young people, future generations, and the other species that share our planet, to point out that the path the world is on, if we stay on it, guarantees that we will push the climate system beyond tipping points.
>
> This is a moral issue, a matter of intergenerational injustice. Because of the inertia and slow response of the climate system, our generation burns most of the fossil fuels and reaps the benefits while future generations bear the costs. We, the older generations and our governments, cannot pretend that we do not understand this situation—we must accept responsibility.[5]

KEY TAKEAWAY POINTS

- The consequences of a warming planet are vast and will ultimately involve virtually all aspects of the planetary systems that support life.
- While there is no serious disagreement among scientists as to the reality of increased carbon in the atmosphere increasing temperatures, there remains uncertainty as to both the timing and the severity of the consequences.
- Some certain effects are increasing numbers and severity of weather events, damage to life in the oceans, rising ocean levels, stress on ecosystems, and public health concerns.
- One of the more serious concerns is "feedback loops" where a warming climate causes changes that enhance the warming.
- The longer we wait to address climate change, the more expensive the consequences, and the more societal and international stresses would be expected.

DISCUSSION QUESTIONS

1. Projecting the future is always difficult. What other possible concerns would you see with an increasing global temperature? Which of the concerns listed do you think would be most easily accommodated, and least likely to cause disruption?
2. What are some of the effects of global warming that may be affecting your geographic area—or could in the near future?

Chapter Fourteen

The Other Guy Did It!

Concern yourself more with accepting responsibility than with assigning blame. Let the possibilities inspire you more than the obstacles discourage you.

—Ralph Marston

The handsome defense attorney stands before the jury, calm and confident that his client will be exonerated. "Yes, it's true, the defendant was found with the victim's blood on him. And his fingerprints are on the knife that was thrust into the victim's back, piercing his heart. But"—he raises his voice for emphasis—"my client cannot have murdered that man because people have been dying of natural causes for thousands of years!"

What?

One might be inclined to think that this lawyer should seek another line of work.

Yet, if we are honest, we can all be tempted to fall into the trap of his reasoning. Climate has been changing because of natural causes for millions of years without any assistance from humans. It stands to reason that the changes we are seeing must also be just another variance in the natural order of things. It's not our fault!

Certainly a dead body, even one with a knife in its back, must be examined carefully to be certain that death was not due to some last-minute natural cause, a heart attack, a stroke, or cancer that finally kicked in right before the knife.

In that same spirit, and in a normal effort to absolve ourselves of responsibility, we need to vigorously examine possible natural causes for changes we are seeing in the earth's weather systems and temperature.

Scientists have identified certain phenomena that have brought about past changes in climate. Most of these are well understood, some are not. Could any of these let us off the hook? These natural events have been under constant scrutiny for the last several decades, and we will have a brief review here.

FIRST POTENTIAL PERPETRATOR—THE SUN

Because the sun is so important to the earth, it seems reasonable to suspect solar activity as the cause of warming or cooling of the planet. In fact, it often has done just that in the past.

Sunspots and Solar Activity

A German named Heinrich Schwabe in the early nineteenth century was convinced there was a planet between Mercury and the sun, named Vulcan no less, and spent some time trying to find it. Alas, no planet. (Sorry, Spock.) What he did find were sunspots, dark areas on the sun's surface that seemed to appear cyclical.

Later observations showed they are magnetic events that occur on roughly eleven-year cycles and are, indeed, associated with increased solar flare activity. These increases in solar activity have been correlated with slight increases in global temperatures—but only small cyclic changes.

We have had cyclic occurrences of cooling that are not completely understood, but solar activity as well as variances in the earth's orbit, and axis tilt (as a result of various gravitational forces lining up), seem to come together in rough patterns and be largely responsible for the temperature changes.

This is not happening now.

Interestingly, since 1975, when we have seen the sharpest increases in global temperature, the sun has had slightly less than average activity. Bottom line, we cannot assume the upward trend in temperature is due to sunspots or solar activity.

Layers of the Atmosphere Warming Differently

Even more convincing of the sun's lack of culpability is the new distribution of heat in layers of the atmosphere. Our atmosphere is a bit like a layer cake without the frosting. We hang out in the troposphere along with everybody else—clouds, weather, airplanes, and Mount Everest. It is here where carbon dioxide and other greenhouse gases collect. The next layer up is called the "stratosphere." If the warming was due to the increased solar activity, there should be warming at all levels.

In fact, the stratosphere is cooling, which is exactly what we would expect if the warming was due to trapped CO_2 in the troposphere. As the heat builds up in the bottom layer, the next layer up does not get its "fair share" of reflected infrared radiation, and cools. This is what is being measured.

Fluctuation of Warming

One more point. If the sun was the bad guy here, we would expect to see increased warming during the day and summer. In fact, it is the other way around. There is a larger increase in temperatures at night and during the winter, which is consistent with climate change models.

Conclusion—we can't blame the sun.

WHAT ABOUT EARTH BEHAVIOR?

Dansgaard-Oeschger

Another natural cycle, which is not well understood, is the Dansgaard-Oeschger events, which scientists have identified from ice core data. (Extra points for being able to pronounce it.) These seem to happen about every fifteen hundred years and bring about a fairly fast warming of the Northern Hemisphere. This looks like a possible culprit, except for a couple of details.

First, the last D-O event occurred about a thousand years ago. So we aren't due for another five centuries. More important, however, is that D-O is a redistribution of temperature, not an overall increase in global heat. The Northern Hemisphere gets warmer; the Southern gets correspondingly cooler. We are experiencing a warming of the *entire* planet.

Once again, no dice.

Wobbles

What about the wobble of the earth? Yes, it does wobble, moving one way and then back again on its axis by a few degrees, every year and a half or so. This is primarily due to tides and the gravitational pull of the moon and the sun. Here again, any change in temperature or climate on the part of the earth tilting a bit more toward the sun is offset by reverse changes on the part tilting away. This is not an overall global warming like we are seeing.

Interestingly, there is some strong speculation that the wobble may be increasing a little *because* of climate change. It is being suggested that the melting of the ice at the poles and redistributing of water may be changing the wobble.

Our age-old wobble cannot account for the global warming we are seeing.

CARBON DIOXIDE FROM SOURCES OTHER THAN HUMAN FOSSIL FUEL USE

Since the bulk of the carbon on the planet is held in rocks and inorganic materials, this might theoretically have some merit. In fact, part of the carbon cycle deals with the slow movement of carbon in and out of rocks, the earth's crust, and from within the earth. That cycle involves tens of thousands of years. But could it be that the CO_2 we are seeing is from unnoticed volcanic eruptions, or large rock deterioration, or gases somehow seeping up through the crust of the planet? Maybe it is coming in from asteroids?

This won't fly. Human fingerprints are all over the increasing amounts of CO_2 we are seeing in the atmosphere. One of the most important fingerprints is from the study of carbon itself. This is a little tricky.

Carbon comes basically in three flavors (called "isotopes"), depending upon how many neutrons an atom has in its nucleus. You remember (of course you do) from chapter 5 that all carbon atoms have 6 protons. The majority of carbon atoms also have 6 neutrons, and we call these atoms "carbon-12" (^{12}C). There are two other types, "carbon-13" (^{13}C) with 7 neutrons and "carbon-14" (^{14}C) with 8 neutrons. Most of us have heard about ^{14}C because it is radioactive and is used in dating organic materials.

This is a story about ^{12}C and ^{13}C. Plants prefer the lighter ^{12}C to ^{13}C. Therefore, they have a larger proportion of ^{12}C in their structures than is found overall in nature. So do you, since the carbon in your body came from plants. And—here is the kicker—so do fossil fuels because that carbon also came through plants, long ago.

And . . . so does the carbon dioxide that we are measuring in the air. As the CO_2 increases in the atmosphere, the ratio of ^{12}C to ^{13}C increases, demonstrating without a doubt that the increases in atmospheric carbon came originally from living matter, which means from fossil fuels.

Busted. This is not some sort of rock off-gassing, volcanic eruptions, or contributions from some celestial source. This is seen as the real smoking gun (or bloody knife if you prefer).

We can further verify the source by crunching the numbers. We know how much fossil fuel we have been burning since the mid-nineteenth century, and how long it stays (several hundred years or more). It all adds up. The amount of increased CO_2 in the atmosphere corresponds to the increased CO_2 we have been putting there.

We have our fingerprints all over this excessive carbon dioxide.

WHAT CAN THE DISTANT PAST TELL US ABOUT HIGHER CO_2 LEVELS?

We have evidence going back eight hundred thousand years, from ice cores and other sources, that the global temperature for that period has remained within about a 1°C range and CO_2 concentrations between 180 and 300 ppm. But further back, in the age of the dinosaurs, there is evidence that the temperatures were quite a bit warmer and CO_2 could have been as high as 1,000 ppm.

Several things become apparent. This level of carbon dioxide, long before the Industrial Revolution, can scarcely be blamed on the human species, and is considered to be a result of heavy volcanic and other geological activity. This warming, however, was very slow in coming—over tens of thousands, maybe millions, of years, not the rapid warming we are seeing today. Plants and animals survived and did well because they had the ability to evolve and adapt to slowly changing conditions.

At these temperatures, there were no ice caps. Sea levels during portions of the dinosaur period were more than two hundred feet higher than today, which would drown all coastal areas and low-lying areas including Florida and Texas—anything currently at less than two hundred feet above sea level.

Just because we are pretty sure that some types of life survived, and did well, that warm period is of little comfort to us and modern civilization.

The only rapid warming like we are seeing right now, that we know of, is called the "Permian mass extinction," 251.9 million years ago. This was triggered by a rare event called a "large igneous province," massive volcanic eruptions in Siberia that emitted huge amounts of carbon dioxide and methane at speeds similar to what humans are emitting now.

This compares to the speed of warming we are currently seeing. Being natural, however, doesn't mean it was benevolent; it caused extinction of around 90 percent of all species, quite a bit worse than the event that wiped out the dinosaurs. The extinctions of ecosystems continued for about sixty-one thousand years, and it took more than ten million years for the planet to recover.

Imitating this kind of natural occurrence with anthropogenic dumping of massive amounts of GHGs in the atmosphere is not in our best interests. Our current worldwide warming may seem slow to a human, but in geological time this is the climate equivalent of the Daytona 500.

NATURAL FACTORS THAT SCIENTISTS HAVE NOT YET IDENTIFIED

Science is never finished and must always be open to the possibility of a heretofore unidentified natural event that may be partially responsible for the warming we are seeing. However, as we understand more and more about the changes in our atmosphere, the importance of this hypothetical natural cause becomes vanishingly small. The effects we see line up with the increases in CO_2 and other GHG emissions we are measuring. Any unrecognized enhancing or mitigating natural process is playing only a very small part.

The verdict is in. Beyond a reasonable doubt, Mother Nature is not the perpetrator. In the words of Walt Kelly's *Pogo*: "We have met the enemy and he is us."

KEY TAKEAWAY POINTS

- The climate of the earth has frequently changed from natural causes. Natural climate change has been slow, over thousands or even millions of years, unlike the rapid temperature (in geological terms) we are seeing now.
- None of the known natural causes of the earth's changing climate can account for the realities we are seeing in today's global warming.
- Human fossil fuel emissions of CO_2 is the only explanation that is consistent with our current global situation.

DISCUSSION QUESTIONS

1. What natural geological activity can you imagine that might explain current conditions including higher carbon dioxide concentrations and increased temperatures? How could this activity be studied?
2. If global warming was demonstrated to be due to natural causes, how would that change the human response to it? What difference would it make in terms of our future or our use of fossil fuels?

Chapter Fifteen

The Trouble with Scientists

Valid criticism does you a favor.

—Carl Sagan

What we know about global warming, climate change, and the energy crisis comes from the slow, tedious, decades-long work of thousands of scientists worldwide. But sometimes they have a bit of an image problem.

The popular depiction is a nerdy sort of person in a white coat with a pocket protector, glasses, and, of course, a lousy haircut. They frequently have a crazed look in their eyes, seem fascinated with blowing things up, and tend to live in their own ethereal world, concentrating on details and precision that fatigues the rest of us. At least this is what we see in the movies. On something this serious, why should we throw our lot in with these folks?

In fact, people who go into science are pretty much like the rest of us and share the full gambit of characteristics. Most of them rarely blow things up. As with all stereotypes, however, there is a certain kernel of truth to these popular perceptions.

THEY CAN BE LOUSY COMMUNICATORS

A few do a decent job. Bill Nye the Science Guy is fun to watch and Neil deGrasse Tyson explains well, but as a rule, PR is not their thing. Judging by the number of Einstein posters that grace dorm rooms worldwide, everybody loves the guy. He had lots of fans and admirers. But listening to his explanation of the theory of general relativity was a sure cure for insomnia.

Marc Moreno, a man who is honest enough to admit he works to stop action on climate change, points out: "You go up against a scientist, most of them are going to be in their own little policy world or area of expertise, very arcane, very hard to understand, very hard to explain and they are booooor-ing."[1] It's true, scientists do tend to lose their audience in an avalanche of details and don't fare well when up against well-paid, professional spokes-people.

THEY EQUIVOCATE

Sometimes it seems that scientists can be devilishly hard to pin down. Ask if Hurricane Sandy or a wildfire in Texas is a result of global warming and you can get definite answers from politicians, media pundits, religious leaders, and the man on the street. Everyone seems to know "for sure" that these events were, or were not, a result of CO_2 emissions.

Ask the same question of scientists and they will pull out mountains of research, copious studies, computer models, data from satellites, decades of ocean observations, ice cores, sediment cores, and statistics going far back in time. Then they will respond with "These events *are consistent* with global warming" or "There is a high level of certainty," or, everyone's favorite, "Maybe."

You want to grab them by their starched white lapels, whack them on the head, and scream "Yes or no?! Is it a result of global warming or isn't it?!"

Why so cagey?

THEY ARE FREQUENTLY WRONG, AND THEY ADMIT IT!

Throughout history, scientists have been wrong—sometimes spectacularly so. Aristotle was wrong when he said heavy objects fell faster than lighter ones. Ancient biologists got it wrong when they said that new human life came only from the man—the woman was just a vessel for the man's progeny.

A century ago physicists were wrong when they said that the universe was stagnant, and a few decades ago they were wrong when they thought that the expanding universe was slowing down. In fact, it is speeding up and so far, there is no solid explanation why.

In the nineteenth century many scientists felt that the universe must be filled with a rigid, invisible material that they named "luminiferous ether." This assumption came from the fact that light waves moved through space. Given what they understood about waves, something must be "waving." Many disagreed and felt the concept awkward, but couldn't come up with a better explanation for the movement of light.

In 1881 a German-born American physicist by the name of Albert Michelson set out to end the controversy and prove the existence of this ether once and for all. He designed a very clever, complicated apparatus called an "interferometer." Imagine his disappointment when all his hard work demonstrated just the opposite—empty space is empty space. No ether. He, and the majority of his colleagues, had been wrong and were left with no explanation as to how light waves could move through space.

There is no question—science, the study of the natural world, often gets it wrong, at least initially.

So why should we trust these folks with something as important as the future of our civilization? Because each of these "weaknesses" is actually what makes scientists and the work of science powerful and reliable.

FIRST—THAT LOUSY COMMUNICATION

Poor communication skills generally indicate one who has not spent time learning to communicate, but rather becoming highly versed in one's field of study.

True experts in all fields, whether it is hairdressing, yoga, auto mechanics, or financial investment, have their own language and are frequently not good at explaining what they do to us novices. However, their inability to extract themselves from their terminology and the details of their profession doesn't negate their expertise.

Scientists have an added problem in that their world is filled with numbers, equations, graphs, and statistics that seem foreign or dull to those of us who have not spent a lifetime learning about these things.

Further, much of the hard work and excitement of science is found in the details. It can be difficult for scientists to stick to the big picture when the little parts are so cool.

Just as you are unlikely to reject your favorite hairdresser because of an inability to verbalize the finer points of his or her art, scientists need to be judged on their accuracy and attention to detail, not on their entertaining and clever presentations—or lack thereof.

When looking for accurate information, siding with the person who is the best speaker, or who has the wittiest retorts, or who knows to remember to turn on the microphone, is probably ill advised. We need truth based on verifiable evidence.

By that criteria, scientists win by a mile.

WHAT ABOUT THIS AVOIDANCE OF ABSOLUTE ANSWERS?

Care and avoidance of hyperbole is not a weakness, but an important strength. It is what makes science so valuable, and what ultimately gets to the objective truth.

Science is evidence based; this means conclusions must be supported by verifiable data. One of the most serious breaches of scientific integrity is to claim a position without support from observation—or to overstate beyond what the evidence shows.

It's not possible to say that any one extreme weather occurrence is a result of global warming because there are not enough facts about a single event. (Although more and more it looks as though most modern weather events are at least influenced by the changing climate.)

There have been extreme weather incidents for as long as data has been collected. What can be said, without a doubt, is that these extreme occurrences are happening more frequently, and are often more intense, and that is exactly what is predicted with global warming. This a scientist will state with confidence because there is so much data to back it up.

Anybody with half a throwing arm can toss a snowball on the Senate floor and claim it is proof of global cooling, or global warming, or that the Mets will win the pennant, or, for that matter, that the Kardashian marriage is in trouble. If you are not dependent on evidence and facts, or bound to anything but your opinions, the sky is the limit.

To come to the truth, one must be willing to entertain all possibilities that are consistent with the evidence. Fact-based scientific research can be slow, tedious, and often boring. Accumulation and evaluation of data and information and establishment of an overarching explanation takes time, and a lot of people.

Painful as it can be, all conclusions are always subject to further information that might throw your best ideas into question. In the scientific world there is no truth so sacred, and that includes our current global warming realities, that they are not subject to reinterpretation when new verifiable evidence is presented.

However, as data accumulates, the possibilities narrow down and the scientific community may come to consensus. Sometimes this is relatively fast but usually the process takes years or decades.

In terms of global warming, current evidence allows a range of views as to the speed of the temperature rise and the severity and nature of consequences. But the reality of global warming, its threat to humanity, and the fact that it is human caused, is "settled" science among the thousands of experts, worldwide, in the field. It has been for quite a while. The data, statistics, and observations accumulated over the last several decades is overwhelming.

The obligation of scientists to not overstate what the evidence shows is one thing that makes scientific conclusions and consensus valuable and reliable.

BUT, BY THEIR OWN ADMISSION, SCIENTISTS CAN BE WRONG. WHY SHOULD THEY BE BELIEVED?

The willingness of science to be wrong is the most powerful tool it has in the search for truth, and what separates it from opinion and belief.

Science is self-correcting. Part of the scientific process is to constantly search for "chinks in the armor," so to speak. The frequently used term "peer review" means that research and conclusions must be published and subjected to evaluation by others in the field. Does it hold up to scrutiny? Can others repeat your work with the same results? Basically, everyone is permitted to take potshots at your efforts. Scientists are born skeptics, questioning and doubting. If your data holds up through this process, you may be on to something.

It also means that concepts that are well accepted are also open to revision. Michelson, as well as much of the scientific community, had to give up the idea of luminiferous ether filling space even though he did such a good job of explaining how light was transferred. His evidence proved this hypothesis to be wrong. This demonstrates the strength of embracing "wrongness."

Michelson's experiment laid the groundwork for the important "particle/wave" theory of matter and the photoelectric effect that makes today's solar panels possible. Enthusiastically searching for misunderstandings and errors is a major step in moving toward important truth.

Can They Ever Be Sure?

One might be inclined to suggest that if scientists must always be open to the possibility of being wrong, how can they be definite of anything? The answer lies in the volume and quality of evidence.

In the mid-twentieth century there were increasing numbers of individuals who were concerned with the possibility of global warming. They knew that carbon dioxide in the atmosphere was how the earth stayed at living temperatures, and it was obvious that we were, even then, adding additional unnatural amounts of CO_2 to our air. It seemed likely that this could increase the greenhouse effect beyond current levels. Interestingly, at the same time, there was a hypothesis that more powerful forces might indicate some cooling of the planet.

There was not enough data to be certain. Nor was there the evidence to project what the consequences of this warming (or cooling) could be.

Fast-forward several decades and the amount of evidence from a variety of sources is staggering. New technologies and a dedication of resources to the problem have clearly shown rising temperatures, delineated consequences, and identified the increase in CO_2 to be from fossil fuels. Today we can even study the climate and CO_2 levels far back in time to further understand our world.

The evidence all points in one direction, and the reality of global warming has emerged to be as definite as anything science has determined, including the elliptical shape of the earth's orbit or the existence of radio waves.

We must be open to the possibility that some new piece of information will prove us all wrong and somehow account for all the changes we are seeing on earth. However, at this point, the evidence is so overwhelming that to ignore it would be like driving your car, with your kids unbelted in the front seat, full speed into a massive stone wall because you think maybe the wall is an illusion. You want to bet your life on it?

Vigorously searching out and accepting where one is wrong is a necessary step in achieving truth and reality. It is also, paradoxically, the best way to be certain of conclusions. The scientific process does this well.

MAKING A NAME FOR ONESELF IN SCIENCE

Do scientists need to stick to the approved "dialogue" to progress in their field? Actually, in honest science, that could not be further from the truth. As Neil deGrasse Tyson points out, "Boldly going where hundreds have gone before does not make headlines."[2]

It is the nature of science to be searching for truth, in whatever form it arrives; verifying established principles certainly plays a part. But the thrill of science, and recognition, comes from moving beyond standard ideas to something new.

The Actual Road to Fame (If Not Fortune)

We have heard of Kepler, not because he "went along" but because he, outrageously, demonstrated that the solar system did not revolve around the earth, but the sun. Galileo is in our science books for establishing a scientific method and proving Aristotle wrong (and the heretic Kepler right). Newton shows up because he had the audacity to buck common ideas and demonstrate that forces work the same everywhere and are based on mass.

The young Einstein would probably have spent his life working in that patent office had he not proved Newton wrong. Darwin, of course, is still vilified by many for breaking with past understandings, but that is why his is a household name today.

Right now, there could be no bigger coup in science than for someone to demonstrate that global warming is, in fact, not happening. It's all been a big mistake. This person would be greeted with fame and prestigious awards.

But here is the catch. There must be legitimate, verifiable evidence. Opinion, prejudice, wishful thinking, or a really odd hairstyle will not do the trick.

So There Are No Dishonest Scientists?

In our modern world there are scientists employed by industries and organizations who must put the needs of their employers ahead of true science. These individuals may not be able to be totally forthcoming with their research.

One dramatic example was the fact that researchers within the tobacco companies had determined many of the harmful effects of smoking even as industry spokespersons insisted that there was no link between tobacco and disease. For the sake of their employment, these researchers kept quiet about their findings (although the break in the stalemate came when internal research was leaked—by whom we still do not know).

Just as troubling, and perhaps more discouraging, are certain scientists who "sell out" their credentials to those who have a vested interest in suppressing actual scientific truth. The book *Merchants of Doubt* by Naomi Oreskes and Erik M. Conway documents this troubling phenomenon. There is a movie by the same name.

All professions have their unscrupulous members. This underlines the importance of looking at sources and at least questioning those representing a vested interest, ideology, or profit-making industry, rather than actual empirical truth. Looking for consensus among many scientists as opposed to putting faith in a small group of spokespeople is wise. But there is enough deception out there, not just in global warming, but in health claims, and just about anywhere money or fame is to be found, that we can all be forgiven for getting a bit mixed up.

Individuals who pursue science do it because they have a need to figure out how things tick. Legitimate scientists basically like to solve puzzles; that is their nature. Twisting or trimming a piece to make it look like it fits, when it doesn't, simply is not satisfying.

The near-universal agreement by eminent scientists all over the world about the seriousness of our climate change is not because they are afraid to show information to the contrary. It is because all the evidence taken from extensive and varied experiments and observations points in a single direction: human-caused global warming that must be addressed while there is still time.

KEY TAKEAWAY POINTS

- Scientists are not always the best communicators. They spend their time and effort in the pursuit of empirical truth and not in honing their presentation skills. This does not take away from the validity of their work; rather, it indicates more of a devotion to it.

- To maintain the integrity of scientific research it is a basic tenet among scientists that positions taken must not be overstated, and must always be backed up by evidence.
- The fact that science is self-correcting, that is, that there is vigorous effort put into finding and correcting errors, is one of the most powerful parts of current scientific research and exploration.

DISCUSSION QUESTIONS

1. Can you think of a time when you "overstated your case" only to find out you were wrong?
2. If one is searching for the truth, is it accurate to say that one must always be open to new evidence and the possibility that previous conclusions may have been wrong?

Part IV

How Did We Get to Here?

Chapter Sixteen

From the Dawn of Humanity until the Industrial Revolution

Fire made us human, fossil fuels made us modern, but now we need a new fire that makes us safe, secure, healthy and durable.

—Amory Lovins

The week before Christmas in 1994, three explorers in southern France came upon a cool updraft of air, a likely indication of an underground cave. After extensive effort moving rocks to create an opening, the smallest of the three, Eliette Brunel-Deschamps, squeezed in. She was followed by her companions Christian Hillaire and Jean-Marie Chauvet. Descending deep in the darkness, they eventually came upon a series of large caverns. Eliette cast her light on a wall and famously shouted, "They have been here!" Chauvet is quoted as saying, "I thought we were dreaming."[1]

What they had discovered was a huge cave about nine hundred meters (three thousand feet) in length, unvisited for tens of thousands of years. The walls displayed hundreds of ancient paintings of animals, some of which are extinct today. Dated, by best estimates, to be thirty to thirty-two thousand years old, they are the oldest cave paintings discovered.

The caves were named after Chauvet, the leader of the group, and continue to be studied for what they can tell us about our Stone Age ancestors and how we became the dominant species on Earth.

BECOMING HUMAN

There is an ongoing debate about what separates human beings from our animal cousins. Opposable thumbs is one suggestion; larger brains, upright stance, use of tools, and language are popular, although none is necessarily definitive. But the making and appreciation of art seems to be unassailable. Other species simply don't do this. The sophistication of the drawings by these Paleolithic people demonstrates that this most human of characteristics goes far back in our history.

We don't know what significance these drawings held for these people, so deep in the earth and difficult to view. Many think it may have been for religious purposes to honor the gods, or perhaps in hope of a good hunt. Maybe they just thought they looked cool.

It has been suggested that the way we view these paintings today, flooded with electric light, does them no justice. They were first painted and viewed by flickering torchlight, in caves that remained mostly in deep shadows. It would have given a certain animation to the animal subjects that is lost with modern lighting.

FIRE

This presents another important criterion that divides us from other animals—our ability to master fire and control it for our convenience. The Chauvet caves are deep, dark, and cold. Obviously the drawings could not have been done without some form of controlled light. Evidence of fires for warmth and crude torches was found in the caves.

Fire for other animals, and for the earliest humans, was something to be feared. Sent by powers on high, it could destroy and kill without reason or mercy. But at some point an early scientist discovered that, under the right conditions, it could be contained to provide warmth. Some sources argue that humans have been able to use and control fire for 125,000 years; some suggest much longer.

When cooking began is not clear, but once fire for warmth was available it likely occurred to an ancestor of Julia Child to put a piece of woolly mammoth, or ancient root, on a stick and warm it up. Gastronomic experimentation was born.

A fire pit provided both warmth and light; the next step was fashioning some sort of handheld torch to allow exploration, and eventually the decoration, of dark places like caves.

Another set of caves has been found on the western side of France. The Lascaux caves also bear remarkable animal drawings. They were painted some ten to fifteen thousand years later than the Chauvet caves. Rather than torches, clay lanterns were found that burned rendered fat. This was a major technological step forward.

FOSSIL FUELS

The use of fossil fuels reaches back in our history as well. We know that peat (the precursor to coal) has been burned for fuel for at least tens of thousands of years.

There are reports of coal being used by the Chinese as early as 1000 BCE; some suggest earlier than that. Coal was probably used by ancient China for heat, as well as a material to carve for decorative purposes. Marco Polo, who visited China in the thirteenth century, described the coal he found there as "black rocks, that burn."[2]

By the beginning of the nineteenth century, coal, particularly in England where it was abundant, was being used more and more along with wood for cooking food, heating, and lighting.

The Chinese refined crude oil and used it in lamps about three thousand years ago and were also pioneers in the use of natural gas. In the West, until the nineteenth century, oil, which was found bubbling out of the ground in a few places, was used mainly for medicinal purposes, seen as a relief for arthritis among other things. Its use as a combustible material was not appreciated.

By today's standards the amount of fossil fuels used globally prior to the nineteenth century was insignificant.

WIND, WATER, AND MUSCLE

Wind and water energy have been used far back in our history for grinding, hammering, sawing. Think Dutch windmills, but they weren't the only ones. Waterwheels in streams are a very old innovation.

Beyond that, work was done by muscle. Fabric was produced by hand spinning, hand looms, hand stitching. Transport, and the tilling of fields, relied on humans and animals. Manufacturing of objects was done by skilled hands, one item at a time.

Then, everything changed.

KEY TAKEAWAY POINTS

- One of the criteria that separates humans from other animals is our ability to control and manipulate energy. Evidence indicates that humans have been using fire for probably 125,000 years and maybe much longer.
- Prior to the nineteenth century, almost all energy used was for warmth, lighting, and cooking, and some manufacturing. Fuels were from biomass, including wood, straw, and animal materials, with limited use of fossil fuels.

- Coal has been used for heating and light for thousands of years. Oil was used by China for light but its potential was not seen in the West until the mid-nineteenth century.

DISCUSSION QUESTIONS

1. What changes do you think likely took place in the evolution of humans as a result of the ability to use and control energy?
2. Why do you think cooked food "caught on," and now is universal for human beings?

Chapter Seventeen

Four Revolutions

> *I am always struck by the fact that human awareness of our place in nature, like so much of modern science, began with the Industrial Revolution.*
>
> —Kenneth R. Miller

Then with breathtaking speed, humanity took a leap of faith into parts unknown, places it could not even imagine. Four rapid-fire revolutions occurred in the nineteenth century, bringing unprecedented changes that continue to affect every aspect of our lives—how we eat, how we work, how we play, what we own, how we travel, how we relate to each other, and the dangers we face. Any one of these revolutions would have increased global energy use dramatically.

THE INDUSTRIAL REVOLUTION: COAL

It all began innocently enough. Coal mining in the 1700s was an onerous occupation. Deep in the ground, it was dirty and dangerous, and wet, as caves flooded and miners had to slog through water.

In 1712 an Englishman named Thomas Newcomen invented a steam-driven pump to remove the water. It ran by using coal to boil water and forcing steam into a cylinder, which moved a piston. Clever, crude, and inefficient—it did the job.

Fifty years later a congenial young man, engineer, and scientific instrument maker was asked to repair one of Newcomen's pumps. Recognizing the inefficiency of the system, he set out to improve upon it.

James Watt ultimately came up with a far more efficient engine that could be used for more than just pumping. By attaching a series of levers to the moving piston, he could make a wheel rotate. Once coal could be effectively used to make things go around, the Industrial Revolution was launched.

By the turn of the century manufacturing was moving out of homes and into large factories filled with steam-powered machinery. All kinds of goods from textiles, to machine parts, to furniture, could be built faster and far cheaper than by hand. Trains, ships, and agricultural tools became powered by steam engines, run on coal.

There were tremendous social changes. The benefits of industrialization were not evenly distributed. New wealth allowed the upper class to build huge mansions, collect fine art, and erect museums and libraries. This privileged social group had always enjoyed prosperity, but now they achieved a new realm of luxury and extravagance. A middle class also saw new prosperity.

However, many people were displaced from their previous agrarian lives, and as they flooded into urban areas looking for work, life became, at best, dreary and often pure misery. Cities became overcrowded; working conditions were often filthy or dangerous and rife with disease. Wages for this working class were low, and even with an entire family working, it was hard to earn a decent living.

Coal was beginning to be used in unprecedented quantities, unimaginable even a few decades earlier. Coal smoke, including particulates and CO_2, was being emitted into the air, darkening cities; a new substance called smog (smoke + fog) hung in the air. London was famous for its "pea-soup" smog—the most serious of which was in 1952 when twelve thousand people are estimated to have died.

A consumer culture, so artfully represented in Andy Warhol's art, was born, beginning in England and Europe and, over time, spreading around the world. The consumption of larger and larger quantities of stuff, made cheap by manufacturing, brought with it a tremendous increase in the use of coal. Innovation in manufacturing continued unabated, with increasing fossil fuels use. Travel became cheaper, more comfortable, more affordable, and more frequently used.

THE SECOND WAVE: OIL

In the mid-1800s a technique was developed to extract kerosene from crude oil. Kerosene rapidly replaced whale oil, which was expensive and didn't burn as well, to light homes. Within a few decades the industry exploded with the main player being John D. Rockefeller's Standard Oil.

Just when it looked like the oil industry was unstoppable, Edison's electric lamp threatened its existence. Within a decade, however, the internal combustion engine that ran off of another refined oil product—gasoline—

gave rise to Henry Ford's mass-produced automobiles. And oil became the fossil fuel of choice, even eclipsing the amount of coal used.

NUMBER THREE: ELECTRIFICATION

Humans have always been fascinated by, and afraid of, lightning. We all know about Benjamin Franklin's foolish experiment with the kite and a key, and he was foolish, he was lucky to live through it. But he did establish that lightning was the same thing as the electricity being observed within various laboratories at the time. Until the middle of the nineteenth century, it was primarily an interesting plaything for the gentrified scientists. Try as they might, they had problems coming up with a practical use for the stuff.

In 1879 Thomas Edison literally brought electricity home with his carbon filament lightbulb. Electricity became useful and practical.

Today we use electricity for everything from cruise ships to toothbrushes, leaf blowers to elevators. We heat, make ice, and light up our cities to the point that many of us rarely see a truly star-filled night. Humans use electricity for unending entertainment and can't imagine a world without a constant supply. Tom Edison was a smart guy, but there is no way he could have foretold how his little lightbulb would affect human culture and society.

The amount of energy each individual uses, because of electricity, has increased logarithmically. But this electrification has come at another substantial price. Electricity is generated around the world, for the most part, from the potential energy contained in coal. Vast quantities of coal are combusted, releasing enormous amounts of chemical potential energy, and equally enormous quantities of CO_2 enter the atmosphere.

The movement to a fossil fuel society has not been cheap. Drilling, mining, transportation—including pipelines, ships, trains, and trucking—refinement, gas stations, and newer enhanced techniques of extraction (fracking, etc.) represent an investment of trillions of dollars at today's rate by both the private and public sectors over the last 150 years.

Industrialization and electrification is now so ubiquitous that it is hard to imagine meeting our needs with things produced any other way.

GERMS—THE FOURTH REVOLUTION

James Watt, in the nineteenth century, had two wives, the first dying in childbirth at a young age. With her he had five children. His second marriage yielded two children. Of his seven children, three lived to adulthood. To lose the majority of one's children, even for the relatively well off, was not unusual; neither was death in childbirth.

There was one more major social transformation to come that, even more than industrialization and electrification, has increased worldwide use of fossil fuels: the germ theory of disease.

Louis Pasteur, in the 1860s, was not the first to suggest the germ theory, but he was the first to convincingly demonstrate it, and in so doing, inspired a health revolution. Now we could fight back against maladies that had plagued mankind since before Lucy walked on two legs. Average life expectancy has doubled since 1850. Some of this increase can be attributed to better distribution of food and the Industrial Revolution itself, but most of it can be credited to effective vaccines, antibiotics, and control of disease vectors that came from an understanding of microbiological causes of disease.

This, however, has also curtailed nature's abilities to keep our population in check. Suddenly we have unprecedented population growth. World population was 1.2 billion in 1850. In 2017, 167 years later, it is 7.4 billion people—six times as many—and it continues to grow.

Further, people in nations who have not enjoyed the benefits of our energy gluttony are lining up to join in. More people wanting the stuff, conveniences, and transportation offered by the Industrial Revolution, means increased demands for energy.

KEY TAKEAWAY POINTS

- The Industrial Revolution in the nineteenth century has forever changed the way of life of humans on this planet, producing far more goods cheaply and offering convenience and comfort.
- Oil, once just used for lighting, has become the main fuel for most forms of transportation, a lot of heating, and production of synthetics.
- Electrification of virtually all the world has further increased domestic and industrial energy use.
- The germ theory of disease has resulted in effective medical treatment and a rapidly increasing population wanting to take advantages of the luxury that the vast use of energy brings.

DISCUSSION QUESTIONS

1. What would daily life have been like prior to the Industrial Revolution and electrification?
2. Spend some time looking around your world. List the various parts of infrastructure that you see that have the single purpose of bringing energy into your life.

3. Is there anything in your past that may have ended your life or that of someone you know, were it not for antibiotics, vaccines, or medical care?

Part V

The Way Forward

Chapter Eighteen

Getting More for Less

Efficiency

We'd find more energy in the attics of American homes (through energy conservation measures) than in all the oil buried in Alaska.

—Amory Lovins

No question, we have ourselves in a pickle. Fossil fuels permeate our way of life; yet the need to move beyond them is urgent. The good news is we can do just that, we have the means, we are developing better technology. The place to start is to be more efficient with our current energy use.

Let's face it, "energy efficiency" is a stodgy term. It smacks of deprivation and conjures up images of living in hovels, driving eight-horsepower cars, and keeping warm in winter by huddling the family around a Sterno can, singing endless renditions of "Kumbaya."

However, the definition of efficiency is "using less while getting the same return." Energy efficiency done correctly can be an important part of mitigating climate change while retaining our current standard of living and even extending it to more people. This is a situation where we can absolutely have our cake and eat it, too. Saving energy offers an important contribution as we move toward a renewable world.

Let's dust off the term and make friends with it.

When the Industrial Revolution began, energy was ridiculously abundant. Coal, of course, seemed everywhere in large quantities, and oil just bubbled out of the ground. Little attention was given to efficiency in its use and we still live with many of those inefficiencies.

Typical internal combustion engines are around 20 percent efficient. Only about 5 percent of the energy that powers incandescent lightbulbs goes to light. And houses, uninsulated with single-pane leaky windows, require large amounts of coal, wood, gas, or oil to maintain comfortable temperatures. Today, modern materials, products, and designs can yield more utility and comfort using less—often much less—energy.

CLOSE TO HOME

Light

Most of us have heard the admonition to change our lightbulbs to LEDs (light-emitting diodes). This is one of those "everybody can do it" things that makes a difference. Considerably more complicated than incandescent bulbs, or even fluorescents, LED lighting has much to recommend it. It uses far less energy, and lamps last longer. LED bulbs have a life of fifty thousand hours compared to twelve hundred for incandescents. One major reason industry, governments, and traffic control are going to LEDs is not just the savings in energy but the reduction in labor to change bulbs.

LED lights cost quite a bit more up front, although the price is expected to fall as more and more are used. But over the lifetime of a bulb it is calculated to cost about one-sixth that of an incandescent bulb—not counting any labor savings from fewer lamp changes.

Interestingly, new as LED lighting is, there is research in progress that could improve incandescent lightbulbs to greater efficiencies than LEDs and may be available within a year or two.

Home Insulation

Until recently, most homes were not built with insulation and energy efficiency in mind. Drafty windows, walls, and roofs that brought the outside in were the norm. In many parts of the United States and other countries, regulations for new construction have established higher standards.

All homes need to have quality double-paned windows. Walls, basements, and especially attics need high levels of insulation. And when appliances are replaced, newer energy-efficient equipment is available. Retrofitting older homes is more expensive than switching out lightbulbs, but it also offers considerable energy savings.

New construction of homes that takes into consideration the direction of windows and "passive solar" can save considerably more energy over conventional construction. Passive solar would mean taking advantage of the different angles of the sun from summer to winter. High summer sun can be blocked out of windows with awnings and eaves, which allow the warmth of the winter sun, closer to the horizon, into the home. Massive stone or cement walls can retain heat in winter or absorb heat in the summer, maintaining a fairly stable temperature and reducing the need for heating and air-conditioning.

While these efforts do require up-front funding, they also bring a higher level of comfort, light, and quiet and eventually pay for themselves in lower energy bills. Of course one also gains the satisfaction of reducing carbon emissions.

How much is saved in greenhouse emissions is difficult to determine. Different climates, different situations would yield different results. Estimates of 10 to 25 percent savings in domestic emissions are suggested.

NONDOMESTIC BUILDINGS AND INDUSTRY

With nondomestic buildings the emphasis is also on efficiency in ventilation and air-conditioning as well as heating and appliances. A savings of 25 percent is often seen. New construction built to high energy standards can provide much greater efficiency than that. Energy-intensive industries, such as iron, steel, and cement manufacturing, are becoming more efficient as waste heat is used and new and correctly sized equipment yields energy savings.

TRANSPORTATION AND CITIES

We do love to move about. A very large proportion of current CO_2 emissions is due to automobiles. The obvious first step is to drive a high-mileage car. Vehicles are becoming more efficient due to improved engines, lighter materials, and government mandates. This is important; for every gallon of gas we use, we put 17 pounds (about 8 kg) of carbon dioxide into our air. Hybrids offer higher mileage and therefore reduce emissions.

Electric cars are becoming more common and most major manufacturers plan to have one or more available by 2020. Because of the relative efficiency of electric motors over internal combustion engines, even plugging your new Tesla into your coal-powered electrical source will offer a reduction in greenhouse gas emissions.

A word of caution. One of the most energy-expensive parts of an automobile is its production. Trading in a car for one with slightly better mileage may not necessarily be the best way to reduce carbon emissions. But when the clunker needs replacing, miles to the gallon should be a major consideration. If you can nurse it along until electric cars become reasonably priced, or more cars can run off of biofuels—even better.

Using public transportation can greatly reduce fossil fuel use. Unfortunately, in most parts of the United States this is not particularly convenient. At one time people in U.S. cities had an assortment of public transportation choices. As the car became popular, the systems aged and, probably with a push from car manufacturers and the oil industry, these trolleys and trains largely met their demise. Now, city streets and freeways are crowded with cars, and people spend much of their day in gridlock; public transportation is looking more attractive.

At this point, it is not an easy move. People want fast, convenient, on-time public transportation. But to build a system that can offer those amenities requires a large ridership—something that will not happen until the system is *already* fast, convenient, and on time. Catch-22. Public transportation will require a significant investment and subsidies by government to build a system that will eventually be used enough to generate a profit or pay for itself. It is unlikely that private industry alone could make this kind of commitment.

European and Asian countries, which never became quite as enamored with cars as the United States, have been able to build their systems gradually in response to constantly increasing ridership. Because they continued to maintain them even as cars became available, ridership did not flag. We do have a few reliable systems in U.S. cities, New York being an example. But for the most part, Americans travel on increasingly crowded highways and streets, possibly listening to podcasts on how to avoid road rage.

For the sake of our comfort, our climate, and our future, we need to be prepared as a people to make the investment in modern public transportation and gain for ourselves the efficient freedom of movement available to citizens in many other countries, but all too rare here.

BICYCLES AND WALKING

And then there are bicycles. Granted, not for everyone, but if you question the viability of bicycles, try checking out Amsterdam. People ride bikes everywhere. In front they have child carriers large enough for several children, and carts behind for a week's worth of groceries. They ride in high heels, or carrying umbrellas, and can get through town faster than in a car. You will never see a healthier group of people than the Dutch. They love their sweets and rich food; yet it is difficult to find an obese person. When you do it is probably an American tourist.

The Dutch ride bicycles because their ancient thoroughfares have been redesigned to make it feasible without risking life and limb. There are efforts to do similar things with some American cities. Finding ways to design urban areas to allow more bicycle riding and walking and reduce the need to leap into an internal combustion contraption for every errand can do more than reduce our carbon footprint. It can make our cities more convenient and pleasant, and help us all to get much-needed exercise.

REBOUND EFFECT

There are concerns with the "rebound" effect in which cheaper energy can cause people to use more. We leave LEDs on a little longer because they cost so little. Insulation has reduced energy costs so we up the thermostat a degree or two. A fast train between San Francisco and Los Angeles encourages us to visit Aunt Sophie more often. This can reduce expected CO_2 savings from efficiency measures.

Nonetheless, it is pretty well accepted that efficiency can make a significant dent in our carbon dioxide emissions before we even make major shifts to renewables. No one is suggesting asceticism, but we need to accept the responsibility to invest in waste reduction.

KEY TAKEAWAY POINTS

- Energy efficiency can significantly reduce our GHG emissions.
- Innovations in lighting, insulation, building design, and transportation technology are reducing energy needs.
- There are several levels of efficiency: personal homes and energy use, industrial use, transportation, and modern urban design.

DISCUSSION QUESTIONS

1. Where do you see inefficient uses of energy in your life? What changes could be made—by you or others?
2. How can a city be designed to reduce reliance on automobiles and yet have all the services, entertainment, shopping, and so forth, expected for modern life?

Chapter Nineteen

Electricity

The Problem and the Future

Electricity is really just organized lightning.

—George Carlin

Electricity has one foot placed squarely in the sloshy carbon-producing technologies of the nineteenth century. The other foot, fortunately, is stepping forward into a new, cleaner future. Paradoxically, electricity, a serious part of the problem, shows the way out.

Probably the only time a major scientific breakthrough has happened in a classroom was in 1827 when Danish science professor Hans Christian Oersted was demonstrating an electrical phenomenon. He turned on the current in a wire, and the needle of a nearby compass moved.

That was a tiny movement for a compass but a giant leap for humanity. What Oersted discovered was the connection between magnetism and electricity. When the compass needle was deflected, it demonstrated that a current-carrying wire was surrounded by a magnetic field. Turn off the current and the field goes away.

Prior to this time, electricity was a curiosity. Interesting, sometimes fun, often a bit shocking. One could store relatively small amounts of it in crude batteries called "voltaic piles" and run some through a wire—as Oersted did. Again, the gentleman scientists' plaything. But as far as usefulness, it pretty much bombed.

Once it was realized that electricity moving through a wire created a magnetic field, the stage was set for the discovery that the opposite was true. A magnet in motion could cause electricity to move through a wire. Energy-carrying current could be generated in large quantities. By rotating a coil of wire within a magnetic field, electrons, carrying energy, would flow through the wire. This device is known as a generator.

WHAT IS ELECTRICITY?

Electricity is basically a pickup and delivery system. Electrons, held loosely by metals, can be loaded with a bit of energy. They are like so many itty-bitty UPS trucks sent through the wires, carrying an energy package and delivering it to your toaster, your lightbulb, your fan, or your computer. There, the energy does some form of work during which it is converted to heat and dissipates into the atmosphere. But the electron itself is not the energy any more than the UPS truck is your birthday present sent from Aunt Gertrude.

Electricity is a very convenient energy transfer system that can take the potential energy from various sources and deliver it in a highly usable form to the comfort of your home or business, or your high-speed train.

HOW IS IT GENERATED?

We create the massive volumes of electric current today using the principle discovered by Oersted. Turbines, in power plants, are essentially huge generators with gargantuan-size coils of wire turned within a magnetic field usually at a rate of 60 times per second. Most are turned by steam heated from the energy of coal, natural gas, or nuclear fission. It is the potential energy from these materials that is converted to the electrical energy carried by electrons.

EXPANSION OF ENERGY USAGE

Once we learned to effectively generate electric current, and Edison invented his lightbulb, electricity use rapidly spread through the country and much of the world. It became considered a basic need. In 1936 the U.S. government went to tremendous expense to extend electricity to the rural parts of the country. At that time, most urban areas had electrical service, but the cost of providing it outside the cities was too great for the private sector to bear. It was seen as so essential to citizens that the government stepped in and did it.

So, within a period of sixty years, energy sent by electric current went from basically nonexistent to servicing almost everyone in the United States. Similar expansion was seen worldwide, and its use has only increased since then.

Now, short of living in a cabin in the far reaches of the Alaskan wilderness, if you reside in the United States, or most places on Earth, you are likely to have a constant supply of energy delivered via electrons along wires attached to convenient, and fashionably discreet, outlets all over your home.

MEETING ELECTRICAL NEEDS WITHOUT THE CARBON DIOXIDE

The great thing about electricity is it doesn't care what kind of energy heats the water that turns the turbines. We have alternatives to fossil fuels that can spin wire within a magnetic field.

Hydroelectricity

The kinetic energy of water can be used to turn the turbines. If you take a tour of the great Hoover Dam, or some other hydroelectric power station, you will notice large, noisy turbines similar to those in fossil fuel plants—but these provide a carbon-free alternative. No water need be heated. These are sort of like waterwheels of old, using the energy of falling water. Unfortunately, that only is practical where large rivers can be dammed, and they can be a little hard on some aquatic ecosystems.

Windmills

Modern windmills, using the same principle of rotating coils of wire in a magnetic field, convert the kinetic energy of wind into electrical energy. They do it directly without trainloads of coal or producing carbon dioxide. But it is the same principle. Wind, of course, is solar energy—a result of uneven heating of air and water—and the resulting movement can produce as much (or more) energy at night as during the day. Today, windmills are springing up all over the world. Even some of the largest oil companies have begun investing in them, thinking that offshore wind turbines may make more sense than more offshore oil derricks.

A company in Spain is developing a very promising wind turbine that, as its founder suggests, looks like a huge piece of asparagus that wobbles. It is called a bladeless windmill and the wind moves it back and forth in a motion to create electrical current. The developers say this particular type of turbine is more efficient and more attractive and reduces a number of the drawbacks of traditional windmills, such as potential danger to birds and bats. As of this writing it is not yet commercially available.

In Iceland, clever people have come up with a wind turbine that looks like a boat propeller wrapped around on itself. Small ones that produce small amounts of electricity for charging phones and the like are already in use in public areas. But large ones could produce serious electricity and offer advantages over traditional windmills, including being less intrusive on landscapes and, again, less dangerous to wildlife.

Concentrated Solar Power (CSP)

Outside most homes in Tibet are what look like large shiny metal satellite dishes with a teapot hanging in the center. This is not some sort of ancient ritual, but a practical way to heat water in this land of thin air and abundant sunshine.

Concentrating solar energy is not all that new. You do it in your car on a hot day and it is used in solar ovens around the world. Solar ovens can be either a foil-lined cardboard structure that concentrates the heat or an insulated foil-lined box. Placed in the sun, you have a buildup of heat sufficient for cooking or heating water.

CSP plants use the same principles. They can be elaborate "dishes" like the Tibetan version, a series of mirrors aimed at a tower, or parabolic reflectors that look a bit like water troughs and concentrate the sun's rays on a tube containing some form of medium, such as oil or even liquid salt. This medium can transfer the solar energy to a water boiler and run a conventional turbine. They have the benefit of producing a very hot medium that can be stored in insulated containers and continue to boil water when the sun is not shining.

There are other ways to produce electricity directly from the sun.

KEY TAKEAWAY POINTS

- Electricity was a curiosity until the late nineteenth century, when we learned how to create current on a large scale and found all manner of ways to use it. A constant supply became seen as a basic need.
- Essentially, electricity is a method of energy transportation by attaching it to electrons and sending it through wires linked to homes, businesses, and individuals.
- Most of the electrical energy generated today comes from fossil fuels that are used to turn turbines.
- Wind, water, and concentrated sunlight can also be used to turn turbines and create electrical current without emitting carbon dioxide.

DISCUSSION QUESTIONS

1. How is the electrical energy in your home and business/school produced? What is the source of energy that was converted to electricity?
2. Name several things that you routinely do during the day that involve no use of electrical energy.
3. How would a typical day in your life be different if you did not have electricity?

Chapter Twenty

Turbine-Free Electricity

Photovoltaics

Solar plus batteries is set to begin a dramatic transformation of human civilization.

—*Bloomberg View*, April 8, 2015

"Out here you need 115 acres for a calf-cow unit," explained an Arizona rancher as he led a tour of his Tucson ranch. He was clarifying why so much space was needed in this desert area to raise cattle. At a generous 0.5 acres per lot, this is roughly the space occupied by 230 homes. A hundred cows and their calves would require 11,500 acres, or enough space for a twenty-six-thousand-household community.

This is a tremendous amount of land for cattle raising. Why do we devote so much space to this purpose? Meat ethics and vegetarianism aside, cattle and beef as well as other grazing animals are an important part of our food supply. We have a deep and meaningful relationship with cattle in this country.

However, beyond the joy of a perfect burger or a good brie, the fundamental reason we eat is for the energy. On that hot summer day, the source of all that energy on the hoof beats down relentlessly.

Collecting solar energy via meat is remarkably inefficient. Raising cattle in an arid desert is even less efficient. According to the USDA, at any one time we have about ninety million cattle in residence in the United States, with about 587 million acres of land devoted to grazing and growing food for meat animals.[1] That's an area about seven times the size of the state of Montana.

What if that amount of land were used, instead, in a more direct method of collecting solar energy? Using solar panels that are roughly 20 percent efficient (by comparison, plants weigh in at about 1 percent), we could capture enough energy from those seven Montanas to generate more than six times the amount used by the entire planet.

Okay, relax. We are not proposing eviction for Bossy, just making the point that we, as a people, have made a choice to dedicate huge amounts of space to a very inefficient method of energy collection. Solar panels can thrive in places your typical bovine does not—like the tops of buildings and houses, over parking lots, lining sidewalks, on top of little lighting fixtures, and even in the arid deserts of Arizona. There is no question that we can find the space.

ELECTRICITY—DIRECT DELIVERY

Albert Einstein is well known for his theory of relativity, a fact most of us know even if we don't quite get it. History intervened and he never received a Nobel Prize for that accomplishment despite the fact that most in the world of physics agree he should have.

However, he was earlier honored with that prize for something called the "photoelectric effect," and, as important as his relativity theories are, it is this discovery that may be more important in preserving our world for the future and getting us out of our climate change conundrum.

His "particle/wave" theory shocked the scientific world by suggesting that light was not just a wave but a stream of particles called "photons." This, incidentally, also explained how light could travel through space. No ether was needed to "wave." Light is both a wave and a stream of particles.

If this seems hard to imagine, don't stress. It is a difficult concept to picture. But countless experiments and observations validate this theory. This particle nature of light is basic to how solar panels (also called "photovoltaic" or PV cells) work. Essentially, a photon of the right size comes sailing in and hits an electron on a piece of silicon, giving it a bit of energy and knocking it off its atom. We have an energized electron running free and it migrates to a metal wire. One photon per electron. A lot of light frees a lot of electrons so we have electric current. Hello toaster.

No drilling, no train transportation, no pipelines, no black gunky stuff, no refineries, no gas stations or storage tanks, no spills or leaks or polluted waterways. No middle men at all. Not necessarily even a complicated grid of wires and assorted hardware. Sunlight to silicon to toaster.

It has, however, taken a while to develop solar panels that work effectively and at a reasonable price. Until the 1950s the photovoltaics fell into the category of "this is cool, but not very useful."

In 1955 the first solar cells were available commercially. They were 1 to 2 percent efficient and carried a price tag of $1,785 per watt. The solar cells to run a 100-watt lightbulb would run about $178,500. A bit steep for the typical consumer. The earliest use was powering satellites, by the U.S. and Soviet space programs.

By the 1960s the efficiencies were nearing 10 percent and prices dropped considerably thanks to the research by both NASA, for space, and Exxon, which needed ways to power offshore oil rigs. Exxon was extracting great quantities of fossil energy but had no effective way to turn on the lights.

The oil embargo in the 1970s stimulated a great interest in alternative energy, and PV technology began to take off. Prices dropped dramatically to $20 a watt. PV cells began showing up not just in industry and homes but railroad crossings, highway communication, and lights in remote places where traditional wiring was not practical. Jimmy Carter put some on the White House.

Today the best cells have about a 25 percent efficiency and the prices continue to plummet due to research into new materials, methods, and economies of scale. Cells are becoming thinner, more durable, and more effective. The cost in early 2017 is between $1 and $3 a watt, which would be about between $5,000 and $15,000 for a typical 5,000-watt rooftop installation.

They are being designed as shingles for houses, tiles for roads, and fabric that can side a building. Clever ways are being developed that let photovoltaics inhabit places that others fear to go. Arid deserts, mountainsides, little garden lights, and shade-providing apparatus for outdoor activities and parking lots are increasingly supporting the technology.

Recently, someone came out with a large "flowerlike" PV unit intended for a single home. It looks like a modernistic mechanical daisy, about fifteen feet high, with solar panels for petals. The panels track the sun across the sky, getting maximum exposure. At night, when it rains, or if a hurricane threatens, the petals stack on top of each other and the whole thing folds up and tucks into some sort of container. Quite a conversation piece, but also fairly efficient.

One can only imagine the amount of energy that could be accessible, and the fossil fuels saved, if homes and buildings routinely had solar panels. Hopefully, solar shingles will soon be standard for all new homes. While this would likely raise the price of homes, over a few years the technology would pay for itself.

Photovoltaic panels are a technology whose time has come. Einstein would be proud.

THAT INTERMITTENCY PROBLEM: STORAGE

In 2015 a billboard greeted drivers traveling north on Interstate 70 that showed a powerful-looking man next to the statement "Wind Dies, the Sun Sets. You Need Reliable, Affordable, Clean Coal Energy."

The "clean coal" part is a bit disingenuous, and we will touch on the affordability of coal in chapter 26. But this meme does make a point about wind and sun. Using fossil fuels for our energy has a tremendous advantage over the use of energy directly from the sun. They are vast stores of potential energy. They can be stored as long as needed and combusted at will, nighttime, daytime, cloudy times, or times with no wind.

In the 1971 movie *The Graduate*, a wise elder takes our young college graduate aside and says, "One word—plastics." Of course, it was good advice for a budding young businessman in the 1970s. The world of plastics was on the cusp of breaking loose. Today, however, the one-word piece of advice to Dustin Hoffman's character would be, without question, "Batteries."

Batteries and various methods of energy storage are where our future lies. The sun relentlessly provides us with astounding quantities of kinetic energy. Energy that can do work, lots of it. And we have the technology to collect it. But the task before us is to convert that into potential or stored energy.

Our world is filled with electronic devices, all running on batteries, usually lithium ion. These are fine for small amounts of stored energy. For solar energy to be used routinely we need inexpensive, safe, environmentally sound storage for large amounts of energy. We need batteries attached to your rooftop PV panels that can save the excess electrical energy for a rainy day—literally. Or for nighttime. Even larger storage is needed for solar installations that will run typical turbines and for large solar collection operations.

Be not dismayed, help is on the way. This is another area where progress is being made across the globe. At the same time a greater commitment to research and development is needed.

The Chinese, who have made a commitment to solar, have a very large battery-testing operation that is finding efficient ways to use conventional chemical batteries for storing solar energy. At the same time new battery technology is being developed that uses nontoxic, cheap, and easily available materials like salt water and carbon compounds.

Other types of energy storage are also being developed. In one instance, huge metal disks weighing tons are set to spinning using renewable energy. Then as they slow down they give off that energy, which can be converted to electricity.

Another innovation already in use in a few areas is a two-reservoir system. During the day, water is pumped uphill from a lower body of water to a higher one, and at night the water flows down to the lower reservoir through hydroelectric turbines. The disadvantage here is finding the right geological setting.

In Germany there is research under way that is using some of the chemical processes of photosynthesis to both capture CO_2 from the air and provide chemical potential energy possibly in various alcohols.

The price for energy storage is coming down steadily, and there is a lot of innovation. It promises to be a large source of jobs in the future.

KEY TAKEAWAY POINTS

- The idea of creating electrical current from sunlight has been around for quite a while, but in the last fifty years commercially available solar panels have been developed. Today the price of this technology is falling steadily and the versatility and efficiency is increasing.
- Battery technology and other methods of energy storage are being developed that can offer on-demand electricity when the sun doesn't shine and the wind doesn't blow.

DISCUSSION QUESTIONS

1. PV cells are being used in tiles, shingles, and fabric. Can you think of other ways these sunlight collectors could be used? What different forms could you imagine?
2. What is a drawback that you can see with collecting solar and wind energy, beyond those mentioned? What is an advantage that fossil fuels have over these energy sources?
3. How many things do you own and use right now that rely on batteries for electric power?

Chapter Twenty-One

What Else?

The use of plant oil as fuel may seem insignificant today. But such products can in time become just as important as kerosene and these coal-tar-products of today.

—Rudolf Diesel, inventor of the diesel engine, which originally ran on peanut oil

Powering an airplane with solar panels could get awkward when flying through a cloud. As effective and cheap as wind, solar, and battery technology is becoming, reduction of our CO_2 emissions quickly enough, while maintaining the mobility and convenience we have become used to, is going to take a broad spectrum of innovation. There is no magic bullet. What other ideas are there to reduce our carbon emissions, and how good are they?

BIOMASS, BIOFUEL, BIOENERGY

A young man in Davis, California, drives a battered old car, works odd jobs, corrects teacher exams to pick up cash, and relies on his wife to support the family. He has two PhDs. He spends long hours growing algae, that kind of slimy stuff one sees in oceans and waterways.

This is one of today's young entrepreneurs hoping his family's sacrifices today will show a profit tomorrow. Like John D. Rockefeller, who was on the leading edge of the last energy revolution, our young man is developing a source of twenty-first-century energy that could prove to be just as revolutionary. He has no illusions of attaining the fabulous wealth of Rockefeller, who, after all, ran a monopolistic empire that controlled 90 percent of oil refining at one time. But even in today's well-diversified renewable energy market, he does hope to one day be able to support his family in a business that is good for the world. And he wouldn't mind a new car someday, one that runs on biofuels.

Our algae farmer is developing one of the most promising ways of providing CO_2-neutral energy. He is returning to nature's tried-and-true methodology, photosynthesis. The idea is to skip the ancient stuff and go directly from photosynthesis to oil production to combustion.

Of course, nothing new here. The energy from plants has been used since we first cooked food and wanted to stay warm; there are a number of ways it can be used today including residues from agriculture or forestry.

For some time, there was hope for ethanol, an alcohol produced from corn. The drawbacks to this were significant. First, modern agriculture requires the use of almost as much fossil fuel energy to produce the corn as solar energy gain. Also, it required the use of land appropriate for food, and we need the food. Ethanol is not completely off the table, but more research is needed to find more effective ways to produce it.

Far more promising is the use of the aggressive oil-rich algaes that our young entrepreneur has been working on. These can be grown efficiently in tanks exposed to sunlight. They can be refined to produce oils that can fuel airplanes, trains, and other technology that might be well served by liquid potential energy.

Of course, one can always plant a tree. And tree planting is a good idea; the more natural flora we have on Earth the more photosynthesis. Even more effective would be a moratorium on the cutting down and burning of the world's rain forests and pollution of the seas.

Trees, however, grow slowly. Algae and similar microscopic materials can be grown rapidly enough to meet some of our liquid potential energy needs.

WAVE ENERGY

Many see enormous potential for electrical energy produced by wave action. Australia and Denmark in particular have made great strides in this area. The constant action of waves can keep a submerged generator in motion and, unlike wind and solar, needs no backup batteries. Currently, more research is needed to make it affordable. But it definitely works.

GEOTHERMAL

The earth is hot inside. Through atomic decay (radioactivity) the planet produces about 20 terawatts of energy. The total use by humans is only 17.7 terawatts. However, for the most part, it is not in usable form.

Nonetheless, tapping into that heat can provide clean energy. Iceland and El Salvador already get more than 25 percent of their energy from geothermal. More than 7 percent of California's electricity comes from geothermal plants. Essentially what is required is living near a "hotspot" where the earth's crust is thin enough to allow drilling to access the heat. Cool water can be heated by this internal heat and then used to run turbines.

An advantage is that, like fossil fuels, biomass, and wave energy, geothermal does not require battery technology for storage.

Certain methods of using geothermal warmth do release CO_2. While not nearly what fossil fuels produce, it is a concern. With an investment in research and development, geothermal could be a source of a good deal of clean, renewable, consistent energy.

THE NUCLEAR OPTION

Nuclear power will help provide the electricity that our growing economy needs without increasing emissions. This is truly an environmentally responsible source of energy.

—Michael Burgess

A pound invested in energy efficiency buys seven times more energy solution than a pound invested in nuclear power.

—Zac Goldsmith

Nuclear power is a conundrum. On the one hand it takes far less fuel than a comparable coal or even natural gas plant. A single pound of Uranium-235, which you can carry in your back pocket, provides about the same energy as six million pounds of coal. It can provide large quantities of carbon-free energy that can operate 24/7. What's not to like?

Turns out, a number of things. Everyone is aware of the safety issues. Nuclear power plants actually have a very good safety record, but when something goes bad, it can go very bad. Chernobyl in 1986 had a meltdown caused by human and structural error. Three Mile Island in the United States was not quite as serious, but it could have been, and was also due to a combination of human error and some design flaws. The Fukushima Daiichi in Japan was built to high safety standards. The 2011 disaster was caused by a tsunami, the strength of which no one anticipated or prepared for.

There is danger from sabotage and terrorism. And waste. Nuclear power plants produce waste that will be dangerously radioactive for tens of thousands of years, and so far this problem has not been resolved. Today it still occupies "temporary" storage. It seems that nobody wants it stored in their backyard.

These problems are being addressed. New-generation nuclear reactors are being designed that are far safer. If there is a power outage, or other interruption, these reactors, because they use molten salt rather than water as a coolant, will automatically shut down rather than melt down. Also, a type of nuclear process, currently championed by Bill Gates and his organization TerraPower, would actually use the waste produced by more conventional plants as a fuel. Both of these methods are being heavily researched and offer promise.

Ultimately, however, what might sink nuclear power is market forces. Traditional plants are ridiculously expensive, coming in at about $10 billion a pop. They require significant government assistance in loan guarantees, tax credits, insurance for accidents, and disposal of waste. Some people have estimated that they can cost the government more than the market value of the power they generate. As the other clean energies become less expensive, this might not be the investment that people are willing to make.

Nonetheless, there are those deeply concerned about reducing carbon dioxide in the atmosphere as rapidly as possible who feel we need to bear the costs of nuclear as another bridge to total renewable energy. This is a point where sincere individuals differ.

NATURAL GAS/METHANE

> *Compared to coal, which generates almost half the electricity in the United States, natural gas is indeed a cleaner, less polluting fuel. But compared to, say, solar, it's filthy. And of course there is nothing renewable about natural gas.*
>
> —Jeff Goodell

Many power plants are switching from coal to natural gas, which is primarily methane, largely because of reductions in natural gas costs. This is seen as a positive step in reducing GHG since methane provides more energy for the same amount of CO_2 produced.

However, remember that methane is a potent GHG, about twenty-five times as powerful as CO_2 (although it remains in the atmosphere for a shorter time). Leaks of methane in the processes of extracting and combustion appear to eliminate the many benefits of the gas's efficiency and some suggest it might even be worse for the environment.

If the leaks could be stopped—an expensive prospect—natural gas could be seen as a bridge to a world of renewable fuels. If we are going to reduce global warming, it is important that we set a course for eliminating all emissions of GHG, and it probably makes more sense to invest in carbon-neutral fuels than in new technology to continue the use of natural gas.

KEY TAKEAWAY POINTS

- Several other forms of energy conversion are being promoted and developed that offer promise in reducing carbon dioxide emission reductions.
- Biomass, waves, and geothermal offer the most promise for new carbon-free energy.
- Nuclear power has much to support it and much to be concerned about in terms of cost, safety, and possible sabotage. Improvements in nuclear technology reduce some of the potential drawbacks.
- Natural gas, theoretically more efficient than coal, does not really hold up when gas leaks are considered; reducing or eliminating leaks might make the use of methane financially competitive with renewables.

DISCUSSION QUESTIONS

1. What do you see as the advantages of each of these methods of energy production?
2. What do you see as the disadvantages of each?
3. If you were an investor, which would you choose to invest in, or which would you think your government should invest your tax dollars in?

Chapter Twenty-Two

Just Keep Thinking

The march of invention has clothed mankind with powers of which a century ago the boldest imagination could not have dreamt.

—Henry George

In the early twentieth century a young Italian by the name of Guglielmo Marconi invented the first device that used radio waves for communication—the telegraph. It was a significant breakthrough but originally greeted with a great deal of skepticism. As accomplished and forward-thinking as this young man was, he could hardly have imagined the changes this kind of communication would bring to civilization or the huge variety of direct descendants from his work—including the ubiquitous smartphone—just one hundred years later.

Likewise, it is largely beyond our comprehension to picture just what the technology we develop today will lead to one hundred years hence. We have likely just scratched the surface of the clean energy innovations that will be available in the future. Today's PV panels and wind energy will no doubt be seen as inefficient and cumbersome compared to their future counterparts.

The search for other solutions to our climate change crisis is continuing and it is worth looking at some innovations currently being considered, even those that may not initially appear to offer a great deal of promise. Like radio waves, once considered a clever if useless plaything, it's hard to predict where any of these might lead.

CARBON CAPTURE

Wouldn't it be nice if we had some sort of an apparatus that would just pull the carbon out of the air? It has been pointed out that trees and vegetation do just this; but as important as they are, it is not fast enough to counter the abnormally large amounts already in the air, let alone compensate for the 35 gigatons of carbon dioxide we are emitting every year.

"Scrubbers" for coal-powered plants have been suggested and in fact can work. Essentially, the CO_2 emitted by the plant would be captured and concentrated into some sort of sand- or rock-like material that could then be stored underground. From time to time coal companies have suggested this technology as a way we could continue converting coal energy into electricity without the carbon pollution. It is called "clean coal."

There is one serious difficulty. It is expensive and essentially makes burning coal a money loser. Despite the advertising there are almost no coal-fired plants working with this technology or any planning to do so. Presumably, someone could find a way to make this economically possible, but it would take a great deal of governmental help, which, right now, is in short supply. As we will see later, as the prices for solar and wind continue to come down, investing in this technology seems less likely.

There are, however, efforts to find ways that might actually be able to remove carbon from the air and bury it, or in one case produce a product that could be incorporated as a building material. It is unlikely that it could keep up with the massive amounts of carbon we are putting into the atmosphere, but it could be a useful technology *in addition* to the elimination of CO_2 emissions. Remember that CO_2 can remain in the atmosphere for centuries. That which we have put there will affect our climate for a long time. This kind of technology could, over time, reduce that damage.

Sir Richard Branson, founder and CEO of Virgin Airlines, has offered a \$25 million prize to the individual who comes up with the best way of atmospheric carbon removal. It must be a "commercially viable design" that can potentially remove a target of one billion metric tons of carbon per year. So far no system has qualified, but there are some promising ideas.

SPACE SOLAR PANELS

One difficulty with solar PV panels is that sometimes there are clouds, and it gets dark at night. What if we could put the panels in space where there are no clouds and they could be in an orbit that would have them always facing the sun? Solar energy all the time.

There are, of course, some serious drawbacks. Theoretically, paper-thin solar arrays are possible, could be quite large, and could collect a lot of energy. Getting them up there and sending astronauts up to maintain and repair them might get seriously costly. Getting the energy back to Earth could be done by microwaves, but you probably would not want to be a bird flying through that microwave stream. Some believe these are not insurmountable problems.

The viability of space PV right now seems unlikely because of the falling costs of earthbound solar, wind power, and battery technology. However, given the magnitude of our potential climate change problems, continued research seems prudent.

GEOENGINEERING

Perhaps we could find ways to artificially keep sunlight from reaching the earth, deliberately altering the natural cycles of the planet. One possibility is seeding the atmosphere with sulfur compounds or other chemicals, called aerosols, that would reflect sunlight. This is based on the observation that after very large volcanic eruptions there tends to be a cooling of the atmosphere as particles in the atmosphere reflect sunlight before it reaches the troposphere.

Climate scientists feel that this cure would be worse than the disease. Dumping large amounts of chemicals in our air is likely to have unintended consequences—lots of them. Further, this is not addressing the basic cause of the problem, the increasing emissions of greenhouse gases.

Even if the temperature of the atmosphere were stabilized by a reduction in sunlight, the CO_2 buildup and acidification of the oceans will continue unabated. And this is proving to be possibly the most serious long-range consequence of the buildup of GHG.

Another suggestion has been to add iron to the oceans, which would encourage plant and phytoplankton life. This was tried in a small area and did increase ocean flora, which resulted in both additional oxygen in the water and an increased fish population.

Here again, most biologists and ocean scientists feel this is an idea fraught with danger. The small gains in CO_2 absorption probably do not outweigh the potential damage to our oceans. And, of course, the basic problem of CO_2 emissions remains untouched.

KEY TAKEAWAY POINTS

- It is important to be open to all kinds of ideas and technology that could potentially reduce the concentrations of carbon dioxide in our world. Sometimes strange ideas yield positive results. At the same time, it is essential to work with nature and not against it.
- Some of the technologies being explored are: carbon capture, space solar collection, and geoengineering.

- Of these, carbon capture right now seems to have the most potential, although it would never be able to pull nearly as much carbon out of the air as fossil fuels are putting in.

DISCUSSION QUESTIONS

1. What ideas do you have for making use of the large amounts of clean energy from both the sun and from within the earth that are renewed every day?
2. What do you think might be a good use of geoengineering? What reservations do you have?

Chapter Twenty-Three

The Third Rail Solution

Population

If we do not voluntarily bring population growth under control . . . nature will do it for us in the most brutal way, whether we like it or not.

—Henry W. Kendall

The ill-fated Donner Party was one of hundreds of wagon trains that set out from the eastern United States in search of better life in the West. Because of some bad information and mishaps, these travelers found themselves navigating the treacherous Sierra Nevada a month too late, and were stopped by an early severe snowstorm. George Donner, one of the leaders, had managed to survive the arduous five-month journey including six days crossing the salt desert in Utah, the rigorous climb up the mountains, and even much of the desperate captivity in fifteen feet of snow only to succumb to a microscopic organism. He died of an infected cut on his arm.

George Washington made it through several wars, numerous battles, crossing the Delaware in winter, Benedict Arnold's treason, and eight years as president of a contentious fledgling country, to be finally taken down by a bacterial infection of the epiglottis.

Throughout history there have been plagues that wiped out millions at a time. One began in 541 CE, lasting for more than two hundred years, and killed an estimated one hundred million people in the Mediterranean basin. The so-called Black Death of the Middle Ages caused the death of 60 percent of the population. The potato blight of the mid-nineteenth century killed a million people in Ireland.

Death by infection, infectious disease, plagues, parasites, and famines have been routine throughout human history. In the mid-eighteenth century almost three-fourths of all children born died in their first five years.

It may have been common, but the loss of children, loved ones, friends, and, of course, oneself has always been painful. Nonetheless this has been how nature has kept the human population in balance, which is essential for ultimate survival.

Here again, the nineteenth century brought dramatic change. People began to live longer lives and far more children lived to adulthood. Part of this was due to the Industrial Revolution, which provided better agriculture techniques, food distribution, and living conditions for many. But the overwhelming explanation for the spectacular increase in population that continues to this day is the germ theory of disease and the resulting effective public health.

THE GERM THEORY AND ITS CONSEQUENCES

The Frenchman Louis Pasteur, in the 1860s, demonstrated that microscopic organisms could cause many of the maladies that plagued humans. He began the experimentation that has led to vaccines, antibiotics, and investment in clean water that has so dramatically changed our way of life and our longevity.

Today few of us bury our children and we all feel a bit cheated if we don't reach the age of sixty, seventy, or eighty—a rarity two hundred years ago. We may complain when that "killer cold" makes its way around town, but the plagues of the past, at least for the time being, are history since we have learned about disease microbes, immunity, and public hygiene.

Inevitably, of course, there has been a price. Pasteur and all the early pioneers in disease control were able to relieve tremendous suffering but probably did not envision the long-term negative consequences of their work.

Until the last one hundred years, survival of our species has required having many children to ensure survival of some to adulthood. So we have a dilemma, a biological push to have many children in an increasingly overcrowded world. We have destabilized the natural birth/death cycle of humans by reducing and delaying death. If we want to continue lives without the pain and suffering of early death and disease, we have a responsibility to rebalance that equation by reducing births as well.

POPULATION—THE THIRD RAIL

Population control is not a popular topic; touch it and you are in trouble. Climate control and environmental concerns often hit the front pages, but we mumble or use euphemisms to avoid "that which must not be mentioned"—controlling our numbers. Decisions about bringing children into the world are probably the most deeply intimate and personally important for all humans; understandably we have a visceral reaction to the suggestion that we might not have the personal freedom to make those decisions.

The harsh reality is, if we are serious about solutions to our global problems and our survival as a species, we must address this issue. If we do not work to stabilize the human population, probably everything else we do to bring about climate stability will be overwhelmed by increases in population.

It is that important.

It took some two million years for the human population to reach 700 million in 1750. Two hundred years later, in 1950, our numbers hit 2.55 billion, and in 2017, it is 7.4 billion and climbing. Considering the demands we put on the planet and its resources, this is simply unsustainable and will reach a breaking point, one way or another, in the not too distant future.

The first step is to face the problem, calmly, intelligently, and comprehensively.

It turns out that controlling our numbers can probably be accomplished *without* impinging on anyone's personal freedoms. In fact, there is every indication that the way to do it is to increase, rather than reduce, human freedom.

THE HEAVY-HANDED "DO AS I SAY (NOT NECESSARILY AS I DO)" APPROACH

Humans tend to have a knee-jerk reaction when problems occur and tend to pass some sort of law restricting the behaviors of other people. Sometimes it may work, at least for a while, but human nature being what it is, there is often a backlash and unintended consequences that can be worse than the initial problem.

Such has been the case when governments have tried to control the fertility of their citizens. An example of this is China. Under Mao Zedong, in the 1940s the government restricted birth control and encouraged large families to make up for losses in the war and provide for military and agricultural needs. As a result, China found itself in the 1980s with a runaway population and began to enforce a one-child policy that slowed the increases but brought with it another set of problems that still plague China today. Other official efforts to impose control on population have met with similar results.

Attempting to control how other people live their lives and what their values should be has never been very effective in much of anything, certainly not in trying to control other people's reproduction.

WHAT ACTUALLY DOES WORK

Rather than restricting the reproductive liberty to have children, if the global community, through their governments, makes it possible for women to have all the children they want, paradoxically, it actually reduces fertility.

According to the Guttmacher Institute, a reproductive health research group, there are about eighty-five million unplanned pregnancies per year worldwide.[1] The world's population is growing right now by about seventy-eight million per year. If all women were given the ability to limit unwanted pregnancies, the population could level out. As a general rule, women do not want to have lots of children. What they do want is to have a smaller number of children whom they can properly care for and who will likely reach a healthy adulthood. The way to "control" the population is to allow people to act in their own best interests.

This could be accomplished by providing accessible contraception globally, backed up, in case of contraceptive failure, by abortion access for those who would choose it. Women who want to have additional children for whatever reason would be free to do so. However, those who prefer fewer, or no, children would also have the freedom to make that choice.

A second factor that seems to help delay and reduce the number of children is education. Putting resources into education, especially women's education, also provides women agency to delay parenthood and choose to have fewer children.

This has already been demonstrated to work in the wealthier countries where birth control and education are largely accessible. Citizens are choosing smaller families, and the populations of these nations are stabilizing.

This strategy was signed onto by 179 nations at a UN conference in 1994. But there are still hurdles in attitudes and political will to be mastered. There remain many pressures on people to have larger families than they want. However, increasing reproductive freedom is a win-win situation for all. It can usher in a world where we can sustain our numbers, reduce carbon emissions, and ensure that children are born wanted, with resources for a healthy life.

KEY TAKEAWAY POINTS

- By sparing people the suffering of much disease and early death, germ theory pioneers set in motion a serious population surge that threatens the survival of the human species.
- Human population has increased from 700 million in 1750 to 7.4 billion in 2017 and is approaching unsustainable numbers.

- Human population stabilization does not require heavy-handed restrictions; it can be better accomplished by providing the global population the means to have only the children they want.

DISCUSSION QUESTIONS

1. How is your life impacted by the increasing population of the earth? Do you see examples of it in your daily activities?
2. What pressures do you feel, or do you see affecting others (relatives, friends, social pressure, government, religion, retirement, etc.), to have more or fewer children than you actually would want to have?

Chapter Twenty-Four

Why Are We Slow to Act?

If you believe you can accomplish everything by "cramming" at the eleventh hour, by all means, don't lift a finger now. But you may think twice about beginning to build your ark once it has already started raining.

—Max Brooks, *The Zombie Survival Guide: Complete Protection from the Living Dead*

In early 1980, Harry Randall Truman became a celebrity and folk hero. He was the eighty-three-year-old proprietor of the Spirit Lake Lodge, located on a pristine lake surrounded with staggering views of the snowcapped Mount Saint Helens, ninety-six miles south of Seattle.

There were songs written about him, tourists came to visit, and schoolchildren made banners honoring him; he received several marriage proposals. He spent his last weeks reading fan mail, giving interviews with the media, and caring for his sixteen cats.

His fame resulted from his unwillingness to accept the scientific consensus of an impending, and powerful, Mount Saint Helens eruption. Truman proudly insisted that "this area is heavily timbered, Spirit Lake is in between me and the mountain ain't gonna hurt me. . . . No damn way does that mountain have enough stuff to come my way."

He stood in defiance of the authorities who exhorted him to evacuate the area. It's easy to understand. He had lived in this gorgeous piece of nature for fifty-two years; he loved it and he trusted it to love him back.

On the morning of May 18, Harry, Spirit Lake, his lodge, and his cats were buried under 153 feet of volcanic debris. Mount Saint Helens had exploded with the force of an atomic bomb, devastating an area of twenty-five square miles. The trees he had looked to for protection were stripped and blown down like as many matchsticks.[1]

In many ways Harry is not that unique. We all have a tendency to want to look the other way in the face of inconvenient realities. Something in us admires this independent spirit, willingness to ignore authority, to go it alone.

One reason many of the lifeboats left the *Titanic* half-filled or less, was the reluctance of passengers, and in some cases crew, to really accept the seriousness of the situation. There was no question that the ship was sinking but, at least for a while, it was comforting to ignore the harsh reality, to put off action. A small lifeboat bobbing in frigid waters could hardly compete with warmth, lights, and food. By the time the deck was listing, and denial was no longer an option, the possibility of an orderly escape had passed, and panic ensued. Two of the lifeboats went down with the ship, secure in their davits.

How often do we put off fixing the roof, changing the oil, or the trip to the dentist even with full knowledge that waiting is going to make things more complicated, carry more serious consequences?

It is easy to look around at our comfortable world and ignore the looming crisis. Yes, yes, we know that the ice at the poles is melting, but it doesn't seem to make a difference to me. Bleached coral reefs? Wasn't planning to take up scuba diving anyway. And species die off? Much as we like frogs, their existence doesn't impact us very much—it's not like we eat that many frog legs.

However, this sort of comfortable denial is like a miner deep underground who notices that the canary has died but doesn't let it concern him because he doesn't really like birds.

Right now, we have an awful lot of canaries screaming at us. You would think we would be running as if our hair is on fire. What is going on? At one time we were poised to take action.

OUR FIRST INTRODUCTIONS TO GLOBAL WARMING

In the 1980s, as data from thousands of scientists from around the world came pouring in, the peril we faced became unassailable. Something was up and needed to be addressed.

Leadership was coming on board with the seriousness of the situation. The first world leader to make a statement about it was British prime minister Margaret Thatcher in a speech to the United Nations on November 9, 1989:

> What we are now doing to the world . . . by adding greenhouse gases to the air at an unprecedented rate . . . is new in the experience of the Earth. It is mankind and his activities which are changing the environment of our planet in damaging and dangerous ways.[2]

Not far behind was George H. W. Bush, speaking to the IPCC on February 5, 1990:

> We all know that human activities are changing the atmosphere in unexpected and in unprecedented ways.[3]

In the first decade of the twenty-first century, former vice president Al Gore came out with his slide show/movie called *An Inconvenient Truth*, which delineated the global warming threat and was more successful than anyone had dared to hope. People from all over the political spectrum were speaking out on the seriousness of climate change.

In 2008 a series of public service announcements seated two people of different backgrounds on a couch where they would acknowledge their differences but agree that we must take action on climate change. One of the ads featured Al Sharpton sitting with Pat Robertson. Another with Nancy Pelosi and Newt Gingrich was typical:

> P. Hi, I'm Nancy Pelosi, a lifelong Democrat and Speaker of the House.
>
> G: And I'm Newt Gingrich, lifelong Republican and I used to be speaker.
>
> P: We don't always see eye-to-eye, do we Newt?
>
> G: No, but we do agree, our country must take action to address climate change.
>
> P: We need cleaner forms of energy and we need them fast.
>
> G: If enough of us demand action from our leaders we can spark the innovation we need.
>
> P: Go to *We can solve it .org*. . . .[4]

The country seemed well on the road to bipartisan, broad support for action.

THE PUSH-BACK

And then something happened. There was a major push-back that is still not totally understood. Part of it clearly was a result of "Citizens United"—the 2010 Supreme Court decision that has allowed essentially unlimited money in politics and made politicians and others beholden to wealthy special interests. Well-funded organizations began to spring up, using many of the same tactics that delayed action on cigarette smoking, acid rain, pesticides, and the ozone layer. There were personal attacks on scientists, usually not on the science, which was unassailable, but on the individuals—the messengers

As with so many things in the twenty-first century, the issue suddenly became a partisan one. The parties had to take opposing views. The Republican Party took an official stance that there was "not enough information for action." With remarkable speed, many Republicans of principle, such as Mitt Romney, who had for some time been an enthusiastic supporter of emission controls, toed the official line and became "unsure" about climate change. Newt Gingrich, previously an advocate for taking action, likewise became "unsure," preferring to "wait and see."

As Neil deGrasse Tyson puts it: "Climate change has taken on political dimensions . . . that's odd because I don't see people choosing sides over $E=MC^2$ or other fundamental facts of science."[5]

Was there a change in the science? Were there studies that called previous conclusions into question? No. Quite the contrary. Accumulating scientific evidence left even less room for any doubt. Our world was warming up, our oceans rising; there were increasing extreme weather events, more severe fires; and the potential devastation to the future was becoming clearer. The chemical changes in our atmosphere and oceans were being better understood, and it wasn't good. But our leadership had begun to fiddle.

We would all like the science of climate change to not be true, and we are natural procrastinators. So when someone challenges the integrity of the scientific community, and suggests that science is not to be trusted, many can easily be persuaded against action.

Politicians are even more vulnerable to the temptation of "wait and see" than the rest of us. Al Gore points out that politicians would basically prefer not to act. It is the nature of our political system that no matter what action they take, it can be dissected and twisted and used against them in the next campaign. In our contentious times, it's better to avoid commitment. So the campaign to make climate science appear unsure, not settled, or simply confused is a perfect fit for a timid politician.

IDEOLOGICAL REASONS FOR RESISTING ACTION

One concern that has emerged is an ideological fear that global cooperation on climate change might compromise American sovereignty. There is no question that combating the changing chemistry of our environment is going to have to be addressed on a global scale, and will involve communication, compromises, and mutual respect among nations.

However, this is already happening at an expanding rate in the marketplace, in communication, in entertainment, and in all aspects of our lives. For example, you may ride in a Boeing aircraft, built in America. But the parts for that aircraft (like all airplanes and the bulk of all manufacturing) came from countries all over the world including Japan, Italy, Korea, Germany, the United Kingdom, Sweden, and France. Safety regulations for that aircraft are international as are the regulations for the airlines that fly it.

A different type of example involves the thirty-year-old "sarcophagus" around the ruins of the Chernobyl nuclear power plant in Ukraine. This structure, quickly built to contain vast quantities of gamma radiation, is crumbling and threatening a second disaster. It was recently encased in a huge modern "supertomb" that took twenty years to design and build. This project was financed by the international community, and involved ten thousand people from thirty different countries. One of the major U.S. contributions was the massive robotics that will be used within the "tomb" over the next several decades to clean up and effectively encase the highly radioactive wreckage of the plant's meltdown.

Why was this an international effort? Because, like climate change, it is an international problem and continues to require the very best skills and resources available. Radiation in the air does not stop at the borders.

International commerce, regulations, and cooperation are already extending into our lives in all kinds of ways. The age when we could isolate ourselves from the world is long past. To solve our significant global problems, we need to work with our fellow inhabitants of the globe. Those concerned with maintaining American values would better serve the cause by making the United States a strong partner in these agreements, not attempting to avoid them.

There is also fear of giving more power to the U.S. government, increasing its size. The irony is that the longer we fail to act on climate change, the more we will be dependent on governments to deal with the problems and the more expensive it will become. Increasing extreme events, flooding, and droughts all require action that private industry would be unable to perform. Already most people get flood insurance through the government because a private organization could not subject itself to the risks of large floods in the future. Dealing with climate refugees, preventing ocean encroachment on cities, and public health problems are a few of the crises that will increase, and will require government action.

One of the biggest fears is that business will suffer if we change to renewable energy. Certainly some fossil fuel industries will see problems, although in a few instances they have already begun to invest in renewable energy.[6] But addressing the concerns of the business community is important. We will do so in the next chapters.

Global warming is not an American or a foreign problem. It is not a conservative or a progressive one. No political party has more to gain or lose from it; it affects us all.

We need honest debate about the best methods to address the problem and everyone from every political persuasion and perspective needs to join in that discussion. Climate reality is, at the core, a conservative value. Conserve resources, reduce or eliminate waste, protect the status quo for future generations, and preserve or enhance the quality of life for all—that is traditional conservatism.

Some highly respected Republicans and conservatives are beginning to break ranks with party politics and are coming forward to propose action on climate change that is consistent with a conservative point of view. And that kind of courage is what is called for right now.

It is important to look at the business side of climate change.

KEY TAKEAWAY POINTS

- There was enthusiasm for addressing the climate change problems across the political and social spectrum until about 2010, when a movement to discredit it evolved.
- Despite continued powerful scientific evidence about the consequences of climate change, there are a number of forces that have been able to foment doubt and confusion about the science, and slow efforts to reduce carbon emissions.
- Climate change is not a conservative or a liberal issue. All points of view need to be included for viable solutions.

DISCUSSION QUESTIONS

1. What resistances to action on climate change do you see in yourself, in others, in the country as a whole?
2. Why do you think business interests and Republicans would be the most likely, initially, to resist action on climate change? Why would people from these groups be some of the most powerful for taking action?

Part VI

The Business of Energy

Chapter Twenty-Five

Disruptions and Jobs

Change is the law of life. And those who look only to the past or the present are certain to miss the future.

—John F. Kennedy, address in the Assembly Hall at the Paulskirche in Frankfurt, June 26, 1963

Change is disruptive. By and large we don't particularly like disruption. We like smooth sailing, happy children, sunny days with comfortable temperatures, and a car that runs without incident. Disruption can bring discomfort. It is also what moves us forward, enriches life—and, actually, we are quite good at it.

Human history is one of constant change whether it is emigration from Africa, crossing the Bering Strait to establish a foothold in the Americas, or switching from VHS to DVDs. We do this. Sometimes it brings great benefits. Often it is not fair and certain people lose while others gain. This switch to renewable energy will be disruptive, will affect all of us, some more than others.

Many are recognizing the inevitable changes, seeing the opportunity in change, and moving investments away from fossil fuels to renewables. But for many industry leaders it will mean losses—losses to shareholders and losses in personal wealth. One of the concerns we hear a great deal about is not the losses by the wealthy, but the possible lost jobs for coal miners and oil industry employees.

JOBS

Moving away from a fossil fuel economy will see a change in jobs and needed skills. This is something that occurs any time there is a cultural change or new technology replaces old. In truth, coal mining jobs have been disappearing for decades due to automation. The process of underground coal mining has become highly mechanized and new methods of mountain-top removal are being used when feasible. Modern mining is cheaper and less labor intensive, if sometimes more damaging to the environment.

In 1920 there were about 870,000 coal jobs; in 2010 that number had dropped to 110,000. This drop occurred while the amount of coal mined and exported was actually increasing. We are producing nearly twice as much coal as a generation ago and doing it with far fewer workers. So, regardless of what we do in renewable energy, most coal jobs are never coming back.

And, of course, those jobs are being reduced even more as coal is finding less and less of a market. Power plants and other users of large amounts of energy resources are increasingly turning to methane and clean energy, including wind and solar. These trends are increasing.

Bleak as that sounds, the overall jobs picture is getting better. Much better. Jobs in renewable energy are increasing rapidly. Rather than a job killer, renewable energy is bringing in more jobs. About six times as many people are employed in renewable energy jobs as in coal.[1]

Fossil fuels procurement is an old technology. New jobs don't come from maintaining the old, they come from exploring and developing the new. Throughout history, including the Industrial Revolution, great booms come out of pushes into developing new technology.

The World Bank estimates that U.S. wind and solar creates about 13.5 jobs per million dollars of spending, and that building retrofits—energy efficiency—creates 16.7 jobs per million dollars of spending. This is more than three times the 5.2 jobs per million dollars for oil and natural gas and more than two times the 6.9 jobs per million dollars for coal. Worth noting is that mass transit comes in at a whopping 22.3 jobs per million dollars invested.[2] If jobs are our concern, we know where to invest.

These, of course, are numbers. They do not address the upheaval in people's lives. The first energy revolution abandoned many to poverty while generating wealth for others. Hopefully we have the wisdom to not repeat those errors. This seems like a reasonable place for us to use our collective resources through government and smooth the way.

Al Gore, at a Climate Reality training in March 2017, referred to the *Harvard Business Review* conclusions stating that all current U.S. coal workers could be trained in clean energy careers for an estimated $180 million, which in government terms is next to nothing. Indeed, he suggested that even ten times that much is barely a drop in the bucket.[3] For comparison, a single F-35 fighter jet costs $160 million and we plan to buy 2,443 of them. A wall between the United States and our southern neighbors weighs in at between $17 and $35 billion.

Investing in our own citizens is a bargain. Right now, the renewable energy industry is having a hard time finding trained individuals for a lot of the jobs they have. Retraining people for good jobs means they will be paying taxes, probably more than they did before, and not be looking to government services to survive. It will also help ease the way toward a livable world for the future.

The switch to renewable energy is likely to bring far more jobs than those lost. And with forethought and the political will, the pain of the disruption for individuals can be kept to a minimum.

KEY TAKEAWAY POINTS

- An energy revolution from fossil fuels to renewables inevitably means change and disruption. But throughout history humans have dealt effectively with change and usually have come out the better for it.
- One of the most significant concerns is the loss of jobs from the fossil fuel industry. In fact, moving to renewable energy will result in a net increase in jobs.
- With governmental action we can help workers make the needed transition to renewable energy work.

DISCUSSION QUESTIONS

1. How would changing to all renewable energy sources change your day-to-day life?
2. Why would new technologies and industries provide, on average, more jobs than older ones?

Chapter Twenty-Six

"Free Market"

The Cost of Energy

The sooner we get started with alternative energy sources and recognize that fossil fuels makes us less secure as a nation, and more dangerous as a planet, the better off we'll be.

—Lindsey Graham, Republican senator from South Carolina

Just how much does energy cost? Nothing. It's free. Really. Economists tell us that the price of something depends on supply and demand. The demand for energy is very high, but the supply is monumental.

Don't believe it? Put a little tomato plant in a sunny place. Wait a month or two and you will have an ungainly bush a hundred times as big providing energy-filled tomatoes—for free!

Or lie out on the beach some afternoon and get a rip-roaring sunburn—for free. If you live in the right part of the country, you could even get free ancient energy. Dig down far enough in the backyard and you can get a piece of coal to roast a marshmallow. Free. You can even cook your lasagna in a solar oven for free.

One might be tempted to point out that you had to pay for that tomato plant, and some kind of cage to hold it up. The backhoe rented to get to that bit of coal was not cheap. We could argue about whether the Neiman Marcus swimsuit was really needed for the sunburn, but let's not quibble.

All these expenses are not for the energy. Those are the costs of gaining access to the energy and making it usable. And that is what we pay for when we pay our energy bills. Just what those expenses are and who pays them are worth examining as we seek future energy solutions.

Consider wind energy. Those windmills had to be designed, constructed, transported (no easy task there), installed, and hooked up to whatever system they are serving. They need to be maintained and repaired from time to time. However, with all these methods of energy collection, once the infrastructure is in place, energy delivery is free.

Likewise with solar panels. The sunshine is free; the panels are not because a lot went into their design, construction, and installation. Geothermal energy just bursts out of the ground, but containing it in a meaningful way is tricky and costs money.

Fossil fuels are unique because they are so far removed from where they are used. They must be dug or pumped out of the ground in some of the most awkward places such as the middle of oceans. An offshore oil platform with directional drills seeking out oil deposits in bedrock is a marvel of technology. But it isn't cheap.

We are paying for extraction, transportation, and conversion, not energy. Energy may not be scarce, but *fossil fuels* are scarce and difficult and expensive to extract, transport, and convert.

Energy from the sun is a universal; we can all share in it. The dynamism of the planet has caused fossil fuels, particularly oil, to be concentrated in certain areas of the globe. When it is necessary to reach beyond our boundaries to procure these sources of energy, additional and far-reaching costs of all kinds are involved.

SOME COST COMPARISONS

We often hear that renewable energy is "too expensive" and that it cannot compete in a "free market." A decade ago this was true. However, in recent years, the cost of renewables has reduced sharply. The figures in table 26.1 are from the Department of Energy and show the estimated cost of a kilowatt-hour of various types of energy.

Currently, energy from solar voltaic panels is close to the coal and natural gas range. Wind can be as cheap as the cheapest natural gas or very expensive, depending on a number of factors, including the technology used and the location.

But this is a snapshot in time and one that looks backward. Already we have seen solar PV installations at 4 cents a kilowatt-hour, and many other installations are well below the 12.5 cents listed in this chart.

And we are just at the beginning. As technology continues to develop we will continue to see the cost of renewables continue to decrease.

What's more, most of the cost of renewables is primarily in the installation. After that it is just maintenance. Old coal plants are cheap because they are paid off. Once we have a large installed base of renewable conversion,

Relative prices for one kilowatt-hour of electricity, depending on source of energy.

Power Plant Type	*Cost per Kilowatt-Hour*
Coal	9.5–15.9¢
Natural Gas	7–14¢
Nuclear	9.5¢
Wind	7–20¢
Solar PV	12.5¢
Solar Thermal (CSE)	24¢
Geothermal	5¢
Biomass	10¢
Hydro	8¢

and we are collecting our energy from that big fusion plant in the sky, the cost of electricity will get closer to the cost we are paying for the original energy—which is nothing.

SOCIALIZED COSTS OF ENERGY FROM FOSSIL FUELS

But we are missing something. The prices for fossil fuels listed in the table above do not include much of the actual costs. Not even close.

In the year 1900 gasoline-powered automobiles were still a novelty and gasoline was sold through pharmacies. At that time a gallon of gasoline cost sixty-five cents, or the equivalent of $18 today. That cost would be unacceptable to drivers in 2017. It has dropped in large part due to improved technology; enhanced extraction, refining, and distribution systems; and reduced labor costs. Much of that reduction in price has come from government support.

The costs for fossil fuels have been largely "socialized" to keep the prices within reason. Economists call these "externalities," which means they are not included in what you pay. Dr. Jeffrey Bennett points out in his book *A Global Warming Primer* that because it is society itself that pays these costs, often the government directly, it is a form of "socialism." Not exactly "free market."

What are the socialized costs of coal and oil energy? They are numerous, some of which are not easy to quantify. But many are universally agreed upon.

There is the direct subsidizing of fossil fuel industries for exploration and production as well as tax write-offs for expenditures on these activities. Both of these are paid for by taxpayers and are estimated to be about $20 billion per year.

Health costs, including dramatically increasing incidents of asthma among children, due to pollutants from fossil fuels and water, are estimated by the U.S. National Academy of Sciences to be about $120 billion per year.[1] These are direct costs that are paid either by increased medical premiums, or taxes for government-covered costs. This does not cover the less quantifiable costs of lost work, reduced productivity, and general discomfort.

Military costs are large. Much of the action of our military is involved with keeping shipping lanes for oil open, protecting sources of oil, and dealing with the fallout of regimes and dictatorships in oil-rich parts of the world. The U.S. Department of Defense and others have tried to quantify these expenses, but it is difficult to do. Conservative estimates come in at perhaps $100 billion a year, although there are those who could justify a much greater amount, suggesting that much of our military action in the world, as well as foreign policy, stems from maintaining oil supplies.

Other costs can be much larger. Costs from environmental damage of oil spills, strip mines, and water contamination are often not listed in monetary terms but clearly are prices we all pay for the use of fossil fuel products.

Ironically, our dependence on fossil fuels has sometimes put a price on other "free" goods. To the extent that freshwater has been compromised by fossil fuel extraction, and people are forced to purchase bottled water, what was once considered an abundant free commodity now has a price. Those who live in cities highly polluted with smog from coal burning and car exhaust, and purchase masks or air purifiers, are paying for clean air. And that catalytic converter you bought with your car and may have to repair or replace when stolen? That is another price we pay for clean air, as is the cost of smog tests.

Regulations in the United States have vastly cleaned up the air from automobile pollution, but, again, at a price that is not included in what we pay for gasoline, but covered through our taxes, out-of-pocket expenses, or the costs of environmental degradation.

Between 1970 to 2000 it is estimated that payment for imported oil cost the United States about $7 trillion or about $250 billion a year. The uncomfortable fact is that some of that money goes to support totalitarian regimes and terrorism, hatred, and intolerance around the world, which is counter to the expressed democratic principles of our country. This compromise in ideals, while not monetary, is perhaps the dearest price we pay.

Of course, it is not just the United States that, because of its reliance on oil, is pouring wealth into regimes and governments that are opposed to our values. Much of the world succumbs to the same pressures.

Finally, the costs of global warming are increasing and, if current courses continue, are difficult to imagine. So far, intensified "natural" disasters such as hurricanes, wildfires, and droughts have been manageable, if expensive. Certainly these have cost many billions. We know more will be coming.

Getting anything approaching an accurate figure for just how much the "socialized" costs of fossil fuels are is difficult. A conservative estimate would say about $300 billion per year just in accounted costs for health, parts of the military, and subsidies. Milton Copulos, an energy analyst who worked for both the Reagan administration and the conservative Heritage Foundation, calculates that the "hidden" costs of oil alone (not including coal and natural gas) came to $780 billion per year in the United States.[2] In 2015, the International Monetary Fund (IMF) determined the total global value of subsidies for fossil fuels to be at $4 to $5 trillion.[3]

It has been a conscious choice to absorb many of these costs by the government and society. Fossil fuels have brought benefits to humanity. Before the threats of global warming and damage to the environment were known, the "socialization" of many of these costs made it possible for all of us to afford this source of energy.

But when attempting to compare costs of various types of energy it is essential to recognize that the actual costs of fossil fuels are a great deal more than our electric bills or charges at the pump indicate. One of the most effective ways to reduce the tax burden on citizens would be to move away from fossil fuels.

A major difficulty right now is momentum. While the economic benefits will eventually be very real, the up-front costs of installation and construction are significant. As long as fossil fuels continue to be subsidized and are relatively competitive, there is an impetus to avoid the short-term costs of change. It took a century of investment to build and modify our fossil fuel industry to the efficiency it now provides. But, in truth, we don't have a century to move to clean energy if we are going to mitigate the effects of global warming for ourselves and future generations. We need to speed up the process.

In a "free market" all the costs of the energy should be borne by the customer and not subsidized by the government. If that were to happen with fossil fuels, market forces would make the switch to renewables much faster.

KEY TAKEAWAY POINTS

- Energy is abundant; new supplies arrive daily and it is free. What we pay for are research, development, transportation, and putting the energy in a usable form for our electrical devices, our cars, and so forth.
- Today, the cost of renewable energy is often competitive with fossil fuels, and still falling.

- The costs of fossil fuels involve what economist call "externalities," which are costs not covered by the price paid for the fuel, but borne by society as a whole and therefore "socialized" through government and taxes and environmental degradation. This has kept the price of this energy low enough to be accessible to citizens but now keeps it artificially low relative to other forms of energy.

DISCUSSION QUESTIONS

1. If energy is free, what specifically are you paying for when you get your energy from photovoltaic cells? Trace the costs of a gallon of gas; what are you paying for here?
2. Can you list other ways that the costs of providing energy, any kind, are borne by government or society rather than included in the price of the product?

Chapter Twenty-Seven

Looking for the Possible

Some Free Market Solutions

There is an urgent need to stop subsidizing the fossil fuel industry, dramatically reduce wasted energy, and significantly shift our power supplies from oil, coal, and natural gas to wind, solar, geothermal, and other renewable energy sources.

—Bill McKibben

In the 1990s, parts of the country were experiencing serious die-offs of trees and animals. Some buildings and various materials were degrading and needing maintenance sooner than expected. Waterways were becoming acidified, killing aquatic life. The culprit, it turned out, was "acid rain." Rainwater that normally has a pH of 5.6 (slightly acidic due to carbon dioxide dissolved in the water) was being measured at 4.2 to 4.4 (considerably more acidic). This was recognized as a serious environmental threat to plants, animals, lakes, waterways, and human health.

The acidity was found to be a result of chemicals, sulfur dioxide and nitrous oxides, that were being emitted into the atmosphere, being absorbed by the water, and returning to Earth as rain. These chemicals were coming largely from coal burning.

Many possible solutions were offered, including simple government regulations limiting emissions. The most effective of the solutions proposed was a so-called cap and trade scheme. Essentially, the government set a top limit for total sulfur dioxide and nitrous oxides emissions. It began as a fairly generous amount. Then permits were sold or given to chemical-emitting entities. This was the "cap" part of it.

The "trade" side was that one plant could sell its permits to another. Company A could invest in technology to clean these chemicals out of its emissions and then make money by selling its permit to Company B. The second company would then be permitted a larger allowance of chemicals, at a cost, and the totals would still remain below the cap target.

This offered an inducement to companies to invest in the clean technology and further created a market for the chemical cleaning technology. A company could offset the costs by selling its permit once the technology was up and running. Over time the amount of allowable emissions was reduced and more and more companies, following the market forces, invested in the new technology.

The amount of acid rain dropped precipitately. The scheme was highly successful because the industry could adjust to the needs of different plants and organizations more than with a firm regulation. There didn't need to be all the typical exceptions and adjustments by government regulators. And the result was reported to be at least as fast as government regulation.

CAP AND TRADE FOR GREENHOUSE GASES?

Some see this as an excellent market-based solution for carbon emissions as well. Again, total acceptable emissions would be determined, permits issued, and then companies could buy and sell the permits based on how quickly they move into non-carbon-emitting technologies. As with acid rain, the total acceptable CO_2 emissions would be reduced over time, giving power plants time to adjust to the changes.

CARBON TAX OR FEE

Another suggestion is to simply increase the price of fossil fuels to incorporate the socialized costs. While there is plenty of room for debate about how much the hidden government and societal contributions are, there is no argument that it is substantial. The price at the meter or pump is artificially low. If free market principles were used, and prices reflected more accurately the cost of the products, there would be a shift from fossil fuels to renewables with little government regulation.

This is called a "carbon tax," a "carbon fee," or a "carbon fee with dividends," and generally has support from people across the political spectrum, including stalwart environmentalists and conservative businesspeople.

Points of Disagreement

There are two points of disagreement. First, what should the charge be? Theoretically, a fee should be added that would cover some reasonable estimate of all of the socialized costs. Some suggest the fee should also include the cost of recapture of the CO_2. This last idea would make sense only if we had viable methods of reclaiming carbon dioxide from the atmosphere and sequestering it, which we don't at the present time. Perhaps it is worth revisiting in the future.

However, even a fee that just represents the total externalized costs is not realistic. The price of energy would suddenly skyrocket and do serious damage to our economy. Another possibility is a relatively small carbon fee, to be increased over time.

A conservative Republican group called the Climate Leadership Council,[1] which includes James A. Baker, Henry M. Paulson, and George P. Shultz, among others, is proposing such a fee starting at $40 per ton of coal. Fees would rise steadily and predictably over time, which would allow companies to plan effectively. This idea has been popular among economists and climate change activists for many years. There is good agreement that this would reduce fossil fuel use and speed the transition to renewable energy.

The next disagreement concerns what to do with the fees that are collected. There are three basic schools of thought.

First are those who would have the money go into the general fund and reduce taxes proportionally. This typically would help wealthier people who pay a larger share of taxes.

A second proposal is to dedicate the money to climate change innovation and technology development. Alternatively, this money could be designated specifically for environment problems that result from fossil fuels such as cleanup for spills, health care resulting from pollutants, or extreme weather events. It could also be used to fund other social programs.

A third proposal, suggested by the Climate Leadership Council, calls for a "carbon-fee dividend." The idea is to return the money to citizens in the form of a dividend. All citizens would receive the same amount, which is estimated to be about $2,000 annually for a family of four and would grow over time. The more the climate is protected, the greater the payments. Since all people would receive the same dividend, more of the money would go to poorer people than the wealthy and would therefore be an economic stimulus, which makes it basically progressive.

Conservatives like the fact that it will effectively address climate change without being a tax that grows the government. Rather, it is money returned directly to citizens through the Social Security Administration. As Baker has said, "It is a good proposal, it's simple, it's conservative, it's free market, it's limited government."[2]

Such a tax failed a ballot initiative in Washington State, even though most people were in favor of it. The environmentalists were split on whether the money should be invested in social programs or toward a dividend. The lack of willingness to compromise doomed the initiative, which would have been an important step in reducing carbon dioxide pollution.

Fundamentally, it really doesn't matter where the money goes. Even if someone takes it into the backyard and buries it, it would still address more of the actual costs of fossil fuels and begin to provide a level playing field for clean energy. This would be a powerful way to use the market to move to a renewable energy world.

CAP AND TRADE OR CARBON-FEE DIVIDENDS?

Either of these, handled correctly, could significantly reduce carbon-producing energy sources and provide a major boost to the clean energy industry, creating jobs and stimulating the economy. And either could reduce the need for complex government regulations.

The carbon-fee dividend plan is the more popular.

First, it is simpler. It doesn't require the creation and maintenance of markets.

Second, it allows citizens to see something closer to the cost of the fuels they are purchasing. When the costs are manipulated behind the scenes through markets, the actual costs of energy become unintelligible to most people, rather than a fairly straightforward difference in prices.

Jim Hansen, a former NASA scientist, summarized this proposal in a 2012 TED Talk:

> We can solve [climate change] with a simple, honest approach of a gradually rising carbon fee collected from fossil fuel companies and distributed 100 percent electronically every month to all legal residents on a per capita basis, with the government not keeping one dime. Most people would get more in the monthly dividend than they'd pay in increased prices. This fee and dividend would stimulate the economy and innovations, creating millions of jobs. It is the principal requirement for moving us rapidly to a clean energy future.[3]

KEY TAKEAWAY POINTS

- "Cap and trade" is a strategy that has been effective in using market forces to reduce emissions of the chemicals that cause acid rain.
- Carbon fees, using collected money to reduce taxes, to support environmental or climate change innovation, or to return to citizens as dividends, holds a lot of promise for transitioning from fossil fuels to cleaner energy sources.
- There is support for the carbon-fee dividend plan from both progressives and conservatives who recognize the need to address climate change immediately.

DISCUSSION QUESTIONS

1. What ideas do you have for using market forces to move away from carbon-emitting energy sources?
2. What are some downsides you would see to either cap and trade or carbon fees? How would you address them to make a more viable plan?

Chapter Twenty-Eight

Capitalism

Yay or Nay?

I'm a capitalist. I believe in capitalism. But capitalism only works if you have safety nets to deal with people who are naturally left behind and brutalized by it.

—Thomas Friedman

Naomi Klein, in her well-researched and comprehensive book *This Changes Everything: Capitalism vs. the Climate*, makes the case that we cannot address the problems of the environment and global warming within the confines of capitalism. As we look around our world today, it's hard not to have sympathy with her position.

Others who are equally concerned with the warming planet and the future suggest that entrepreneurship and competition are the very things that push much of the innovation that is providing solutions to our climate crisis. Capitalism, they say, is essential to addressing climate change.

Can they both be right?

Capitalism is defined as private ownership of the means of goods and service, run for profit. Like so many things, it comes in various forms and configurations. It is the details that make the difference.

Certainly there is a type of capitalism, based on profits at *any* cost, that can be brutal and can lead to amassing of fortunes for a few and impoverishment of others. A system where care for the environment, compassion for fellow humans, and even basic integrity can be sacrificed on the altar of profits is unlikely to move us into a cleaner, safer, sustainable world.

However, history tells us any system—capitalism, communism, socialism, theism, probably vegetarianism—can be manipulated to serve only the short-term needs of the powerful. An old axiom, often attributed to John Kenneth Galbraith, says that, "Under capitalism, man exploits man. Under communism it's just the opposite."

It is not necessarily the "ism" that determines the outcomes. What makes the difference is basic integrity, a vision of the organization within the whole, and embracing values beyond just amassing money.

Business decisions need to be inclusive of all and made within the largest possible context. Innovators deserve a fair profit for their ideas, their work, and the risks they take. But if some of that profit comes from "socializing" costs onto society, and the degradation of the planet that supports us, it is not true capitalism and it is not sustainable in the long run.

Sam Geil, founder and CEO of the International Green Industry Hall of Fame, business consultant, and avid capitalist, suggests that, done right, capitalism can provide the impetus for innovation and growth, while also protecting the environment and serving the needs of all citizens and the future.

He argues that good business practices require values beyond just profit seeking. To that end he offers his "four elements of business sustainability." The first is integrity. "If an organization is not going to commit to integrity," he says, "I make it clear I do not want to work with it."[1]

Second is a spiritual/social component that recognizes that any person or group is part of the whole. Actions must be based on a consciousness of right and wrong and of responsibility for consequences. They must address purposes beyond oneself.

The environment is an important element in his thinking. "Great societies have come and gone and many of the issues leading to [their demise] include the stewardship of resources."[2] Passing on the costs of business, whether by tax money going to pay for damage, or by the society as a whole living with a degraded environment, is not true capitalism.

The last of his four elements is economic and recognizes responsibility to both the honest pursuit of profits and the position of the business within the economic system of the country and the world. Geil supports partial employee ownership of companies, removing the adversarial relationship between management and labor, and giving the entire spectrum of employees a clear stake in the company. Shareholders, management, and employees all share responsibility and the rewards.

In typical capitalist form, Geil is not a fan of government regulations. He says many regulations are doing the job that business owners themselves should be doing and suggests some sort of ethics standards and disciplinary body for businesspeople, similar to what other professional organizations have.

This advice is really just recommending the behavior that all conscientious people should want to emulate. Many businesses do demonstrate this kind of integrity and conscientiousness. But right now we have a large business culture that needs reminding of its responsibilities to society and the world.

Much of today's capitalism has become what *Time* magazine called "financialization"[3] where manipulating money dominates in both business activities and politics. Short-term profits have become more important than long-term sustainability. Good business practices have often given way to zero-sum gamesmanship where individuals see their success only at the cost of others. This mind-set cannot be expected to come up with long-range solutions to our environmental and social problems.

Currently, as much of our economic system fails to look and plan for the future, the United States is giving up its economic leadership in the world. Others are stepping into the void, including China, the EU, and even Saudi Arabia, with its recent substantial investments in renewable energy.[4]

However, writing off capitalism would weaken our ability to address the climate crisis. Using market forces, such as the carbon-fee dividend, can be a powerful tool in our reduction of greenhouse emissions. What is needed is a cultural change in how we define capitalism.

We need to find ways to retool our economic system so success and growth do not necessitate oppression and environmental degradation. We need to help our business sector lift up its eyes from the quarterly, yearly, or even five-year profit reports and see a more inclusive and powerful vision of its responsibilities and potential. It's time to return the United States to a place of economic leadership in the world.

The principles of successful capitalism should be part of making our economy work. Climate change needs all hands on deck and that includes the very best of the conscientious capitalists.

Unfortunately, as the United States clings to the dirty and inefficient methods of the nineteenth-century Industrial Revolution, China is seizing the future in renewable technology.

KEY TAKEAWAY POINTS

- There is an argument that capitalism, with its current focus on short-term profits, is not an economic system that can provide solutions to our environmental and climate change crisis. Others feel that the dynamism and innovation encouraged by capitalism is exactly what is needed.
- Capitalism comes in different forms. There is no reason why it cannot respond to the greater concerns of the whole and the future.
- We need to recognize that progress and profit are not synonymous with environmental degradation and oppression, and use the creative tools of the free market, such as the carbon-fee dividend, to help move us to renewable energy.

DISCUSSION QUESTIONS

1. With regard to climate change, in what way do you see profit-seeking capitalism as a detriment to solutions?
2. How do you best see market forces used to address our serious pollution problems and preserve the world for future generations?

Part VII

Into the Next Revolution

Chapter Twenty-Nine

What Lies Ahead?

The prediction I can make with the highest confidence is that the most amazing discoveries will be the ones we are not today wise enough to foresee.

—Carl Sagan

Predicting the future is a sport fraught with danger. Remember the fusion-fueled flying cars and hover boards in *Back to the Future II*? And what was that twenty-third-century clunky communication gadget Captain Kirk carried on his hip? It made silly noises and didn't even show movies. Alexander Graham Bell wouldn't have predicted today's smartphone, and Alan Turing would be astounded by the Internet.

Nonetheless, we need to do our best to prepare for what we can see of the future.

ADAPTATION

Even if we stopped all CO_2 emissions tomorrow, humanity is going to need to adapt to a certain level of climate change over the next century. Climate is slow moving (relative to a human life) and with considerable lag time. We know that our grandchildren will be dealing with the carbon we are putting in the atmosphere right now. Temperatures will continue to climb in the atmosphere, in the oceans, and on the land.

The hope, expressed by the IPCC (Intergovernmental Panel on Climate Change), is to limit the increase in temperatures to 2°C (3.6°F). Some consider this unrealistically low; others think it is possible. Even at this level we will be seeing more frequent extreme weather including drought and flooding, and rises in the oceans. Coastal cities will need to prepare; some, such as Miami, Boston, and Venice, have already begun.

We will need to address the changes in agriculture, beef up our public health services, and prepare for the climate refugees who are likely due to lost habitation, available food, and so forth. As mentioned earlier, it will be those in the poorer countries, those who have contributed the least to climate change, who will bear the greatest repercussions. Hopefully we can keep the disruption to a minimum and with foresight, plan for those individuals, towns, and countries that will be most inundated.

THE GOOD NEWS

Serious as the global warming situation is, it isn't like a war where the best any side can hope for is to survive battered and bruised. The solar energy revolution can not only ensure our survival, it can bring with it a world that is cleaner, quieter, safer, and more convenient for all. If we address this problem soon, our descendants can once again look forward to leaving a world for their children a little better than the one they inherited—the universal hope of parenthood.

The coal energy revolution brought vast changes to our world. With the solar energy revolution, we can likewise expect dramatic changes in the world. We don't know what they will be, but based on our understanding of current technology, there are a few predictions about a fossil fuel–free world we can see as reasonable.

HEALTH

The air and water will be cleaner. Many cities around the world suffer from heavy smog that reduces the quality of life. The unseen particulate matter from burning coal, the ozone, the sulfur compounds, the mercury, and the nitrous oxide that we dump into our air take a toll on pulmonary health, particularly that of children. We would expect to see the rates of asthma and pulmonary diseases, which have gone up substantially in recent decades, drop.

Oil spills that have despoiled our land, rivers, and oceans and killed off countless wildlife will be a thing of the past.

GETTING AROUND

Electric cars are efficient and quiet. They are simpler—fewer moving parts to break and need repairs. The energy to run our cars may be from our own rooftop energy source that we simply plug into when we park the car.

Bill Nye suggests that in one hundred years, we will all be able to call from our twenty-second-century smartphone equivalent, and a driverless electric car will promptly show up to take us to the train or for a drive in the countryside. The personal car with all its expenses and frustrations will become a rarity, freeing up that room in the garage for pool tables and in-home theaters.

Without the political push to increase consumption of fuels, we may go in the direction of Asia and Europe and provide efficient trains that run frequently, and on time.

Already, city planning is under way that will develop cities and transportation within cities that can put humans front and center and reduce the need for gridlock by automobile. Fast, frequent, electric-run trolleys and subways can make travel less stressful. Cheap solar energy can make it inexpensive. Neighborhoods are being designed to accommodate more walking and bike riding, healthier alternatives for those who would want to take advantage.

Perfection of biofuels offers real promise for powering air transport in the future.

HOMES AND BUILDINGS

We will live in homes that are warmed and cooled with sunlight. To a large extent this may be from our own twenty-second-century equivalent of PV panels and energy storage—giving people real independence over energy collection and use, and avoiding large outages. Many may still be dependent on centralized energy conversion from solar, wind, or as yet unrecognized technologies.

New building standards will provide well-insulated, more comfortable homes, and new buildings may be oriented to the movement of the sun to provide optimal comfort with less reliance on forced heat and air-conditioning.

Chances are reasonable that we will have some solar-powered mechanical assistance doing things like vacuuming or cleaning the gutters.

ABUNDANCE

Within a hundred years the solar energy revolution will likely provide more energy in a variety of forms that continues to drop in price. Energy will be available to solve other problems such as drought. Currently desalinization of ocean water is done in some areas, but at great expense and it does require significant amounts of fossil fuels. Solar energy can greatly reduce the cost and make more water available for drought-starved areas. Some have suggested that the next great war could be over water, as the population increases and fresh, clean waters are reduced by pollution. Cheap solar energy could help produce more freshwater as well as help keep water clean.

We have reason to believe that as robotics, running on solar energy do much of the physical work of society, we will have more leisure time. It will probably require cooperative policies, through our government, as was necessary in the twentieth century, to ensure that the benefits of this reduced workload are distributed throughout society and not just to the powerful.

One of the concerns that has been expressed by some is that as energy becomes cheaper and more abundant, goods in general will become cheaper and we may find ourselves consuming more stuff. This would not be a particularly good thing. All pollution is not from fossil fuels. Some of it is from materials used in manufacturing, consuming, and postconsumption waste. One hopes that future generations might have the wisdom to seek fulfillment in ways other than additional consumption.

GLOBAL CONFLICT

Since the horrors of the twentieth-century world wars, humanity has been working toward greater cooperation among nations. NATO, the United Nations, and the European Union are a few of the organizations that could hardly have been imagined prior to those wars. The postwar United States receives a good deal of the credit for ushering in an era of cooperation among countries with the Marshall Plan and efforts to reach out in compassion to former enemies.

Today's wars tend to be over resources: the number one resource that we fight over is oil. And to the extent we have demagogues, they usually are funded by oil. Often the violence in the Middle East is described as religious. Certainly, plugging into ancient religious bigotries and rivalries is a way to motivate a populace.

However, a closer look reveals the "religious" side is often window dressing. There is a diversity of religions all over the world. Islam and Christianity are global. But the fighting is in the Middle East, where the oil is. Western countries have a long history of being involved in Middle East politics, not because we think it is our moral duty to the citizens of these countries, but because we want access to their oil.

Going to solar means we can get the energy we need without having to compromise our values to extract it from other people's land. We still may not agree with Middle Eastern governments, or religious practices, but we will stop financially supporting the worst of them and we will have far less need to engage them in war. We would also have more credibility in any effort to help them move toward political systems more in line with U.S. values.

The world has been coming together through IPCC, the Kyoto accords, and most recently, the Paris accords, to find ways to address this global problem. We are learning more about each other, our common future, and the need to cooperate to survive. One of the greatest legacies of the worldwide solar energy revolution is likely to be a more peaceful world and a greater hope for effective means of conflict resolution in the future.

People will always find things to disagree about. But if we can remove oil from the equation, we may be able to continue the progress made to resolve global conflicts with statesmanship and negotiation rather than violence and the blood of our children. As Louise Burfitt-Dons points out, "Global warming causing climate change may be the ultimate issue that unites us all."[1]

This could be the sweetest outcome of all.

KEY TAKEAWAY POINTS

- It is difficult to predict the innovations and technology of the future, but the solar energy revolution would not only avoid the devastating consequences of continued global warming but would provide changes that could make a better world for our descendants.
- Health, transportation, and living conditions could all benefit from abundant energy.
- Perhaps a major legacy the solar revolution could provide to future generations is reduced global conflict both because of reduced fighting over oil and the cooperation developed among nations as they move to eliminate CO_2 emissions.

DISCUSSION QUESTIONS

1. What are some additional benefits you see to having a world run on solar energy rather than on fossil fuels?
2. The possibility of increased consumption and production of goods because of inexpensive energy was mentioned. Can you think of other disadvantages to a solar energy economy rather than a fossil fuel economy?

Chapter Thirty

Ending the Gaian War

Solar power, wind power, the way forward is to collaborate with nature—it's the only way we are going to get to the other end of the 21st century.

—Bjork

Over the last two hundred years, through creativity, rigorous science, cleverness, and a lot of hard work, our species has developed the technology to provide unprecedented comfort, convenience, and nonstop entertainment to the majority of the earth's human inhabitants. Kings and queens of old would positively drool with envy at the way of life that is commonplace today.

Will it be just a flash in the pan? Will it be a spark of affluence that burns brightly for a few generations and then flutters out, leaving the world in darkness and despair? The answer to that question is in the hands of those of us alive today.

This global warming crisis can be a motivation to move ourselves into a cleaner, healthier, safer world, unburdened by the inefficient methods of the Industrial Revolution (that was *so* last century). We can demonstrate that "progress" does not inevitably mean pollution and ruination. We can build an economy based on renewable resources while reducing our reliability on those that are becoming scarce. We can extend the benefits of prosperity to those who have been heretofore left behind.

But before we get to these remarkable current and developing innovations we need to address perhaps our most serious problem. It isn't technology, or stuff, or even energy; it is an attitude we have slipped into that is so pervasive we can hardly see it.

Without intent or malice, humans have become engaged in an escalating combat against an unlikely and largely docile opponent—the earth. We have increasingly seen our home as an inexhaustible provider of resources, and a limitless receptacle for the waste, toxicity, and pollution we create from those resources, including, of course, carbon dioxide. We have ferreted out nature's secrets and used them to bludgeon her.

Our quest for ever-more stuff and our ignorance of how our choices have been affecting our world are putting us on a collision course with disaster. It's been quite a ride, but this epic battle is coming to an end, one way or another.

Those of us who slap "Save the Earth" stickers on our car bumpers miss the point.

Make no mistake—Gaia needs no saving.

She will continue her travels around the sun unimpeded for another two to four billion years with or without us. It is you and I, and our grandchildren, who will lose if our current course goes unaltered.

We don't need to save the snail darter in California because any of us is a great fan of snail darters. Likely we have never seen one and never will. We aren't concerned about polar bears because they look fluffy and soft, or owls because they are cute, or frogs because we have a fetish for frogs. We don't hug trees because they are so huggable, and certainly not because they need hugs.

Working to preserve all the parts of the grand ecosystem in which we live is not altruistic, it is completely and unabashedly self-serving. If you can't work up a personal coziness for phytoplankton of the sea (and who can?), consider that these organisms are providing a lot of the oxygen you breathe and are the basis of the food chain that keeps plant and animal life in the oceans successful. Living oceans are necessary for human survival. As go the phytoplankton, so go we, eventually.

All these "endangered species" that some of us fawn over and others disdain, are, in fact, canaries in our cave. Each demise represents a gradual weakening of the web that supports human life. Biodiversity, the number of interrelating species from bacteria to worms to elephants, is all part of a whole, the members of which depend on each other. And to disregard the needs and importance of this interrelated life system is to put ourselves in combat with the only home we have, or will ever have.

Despite its apparent passivity the earth will prevail, and we, with all our cleverness, are destined to defeat in this unintended but self-made war. Eventually, rich and poor, we will lose to the studied, quiet superiority of the ancient systems of the earth.

Given those odds, perhaps it would be prudent to change strategies and join the winning side.

The great cycles of the earth are the methods by which the planet has remained verdant and supported life for literally billions of years. The task before us is to begin working within those cycles rather than fight them, ignore them, or attempt to alter them for our short-term purposes. That is really what the oft-repeated word "sustainability" means.

Our global warming comes from an unbalancing of the age-old carbon cycle. We are putting out more CO_2 than can be absorbed through photosynthesis. This in turn has disrupted the water cycle, resulting in changing climates.

But all materials in nature are cycled. We need to find ways to put our "stuff" back into the earth's cycles. This is no more difficult than going the other way, but we have not bothered to set our minds to figure out how. We have just assumed that Gaia would do it for us. She won't.

Our pressing concern right now is with energy and ways to harness it without polluting the atmosphere. But that problem is within the context of our general attitude. How do we live in a way that maintains a livable earth?

All major movements thrive on courage. Courage to invest in Watt's newfangled steam engine. Courage to storm the beaches at Normandy, sit in lunchrooms in segregated Alabama, or fly to the moon in largely untested machines. Today, we need the courage to broaden our perspectives and change the relationship we have with our great benefactor, and ultimate victor—mother earth.

We have role models all around us. One is Republican Robert Durden (Bob) Inglis, who was a congressman from a very conservative South Carolina district. He once considered the concept of global warming to be, in his words, "a bunch of hooey." Appointed to the congressional science committee, he took the opportunity to listen to scientists as they showed him how climate data was collected and what the implications were. He realized he could no longer deny the reality of climate change and its influence on the future of his children.

Congressman Inglis chose to speak out about his new understanding of global warming, despite the fact that he feared it would cost him the next election. It did. He was defeated in 2010. Since then he has launched the Energy and Enterprise Initiative, which is a campaign to promote conservative and free-enterprise solutions to climate challenges.

He explains:

> To be in this situation where fossil fuels are imperiling our future and future generations . . . that becomes a moral problem. We are leaving our children and grandchildren a legacy of people who failed to lead, people who, when it became their time to be awakened, they slept. We didn't have enough faith in the future that could be brought about so we just gave up, we couldn't rise to higher things. I don't want to be part of that, I want to say "we did rise to it."[1]

KEY TAKEAWAY POINTS

- We have an opportunity to make a cleaner, healthier world for ourselves and future generations.

- We need to end the fight with the earth, realize it will not be subdued, and get on board with the basic cycles and life-sustaining forces that have served the earth for billions of years. That is the definition of sustainability.

DISCUSSION QUESTIONS

1. Write down five nonfood items you can see. Where will the materials that make up these items eventually end up?
2. What are some activities you do that would be considered "sustainable," that is, fitting in with the cycling of nature? What are some that do not?

Epilogue

Lessons from the Past

The most important question we must ask ourselves is, "Are we being good ancestors?"

—Jonas Salk

On the top of a hill in old Istanbul, overlooking the Bosphorus, the skyline is dominated by a building that has stood for fifteen hundred years. The Hagia Sophia is huge, the size of a shopping center, with a dome that seems to defy gravity, located 192 feet above the floor. Built in the seventh century, it began as a Christian church, was converted to a mosque one thousand years later, and in 1931 became a museum.

One is struck by the immensity of the space, beauty of intricate mosaics, and the reality that this ornate, breathtaking building has survived earthquakes, storms, and a great deal of social unrest for nearly fifteen hundred years. Just how did people—and it is reported to have taken ten thousand people decades—manage to design and produce such an edifice without modern architectural theory and using only basic tools and hard work?

It stands as a magnificent symbol of the creativity, resourcefulness, and determination of human beings.

Of course, it is scarcely alone. The world is full of staggering monuments, works of art, and engineering projects designed and built by people not unlike us, but without the advantages that the modern world takes for granted. The imposing cathedrals of Europe, the Great Wall of China, the Roman aqueducts, the Egyptian pyramids, the Peruvian city of Machu Picchu, and the dikes of Holland, to name a few, serve as witness to the resolve, foresight, and intelligence of the human being.

We can create. We can cooperate. We can solve problems. We can build, and adapt, and learn. We can muster our resources to respond to threats.

The modern world can also boast amazing accomplishments, including skyscrapers that reach to heights unimaginable one hundred years ago. We have been willing to devote enormous resources toward building tunnels under mountains or the sea and developing complicated energy networks of oil refineries and electric grids to meet our ever-increasing wants.

One of our accomplishments occurred in 1969 when Neil Armstrong and Buzz Aldrin walked on the moon. This was not an easy undertaking and it was not without setbacks and misfortunes.

When, in 1961, President Kennedy famously proposed that we would send a man to the moon by "the end of the decade," the personnel at NASA were taken aback. They simply had no idea how they would go about that. They were still trying to reliably get a rocket beyond the launchpad without blowing up.

The stimulus for this adventure had occurred in October 1957. The Soviet Union launched a tiny satellite called Sputnik. After years of convincing ourselves how technologically superior to the Soviets we were, the United States was broadsided. We were not even close to such a feat.

This was not just nerdy intellectual pride. A country that could successfully launch a satellite could also launch a missile—one that carried the newest of modern weapons, a nuclear bomb. Demonstrating one's technological prowess was to become an important battlefront of the Cold War. This was a threat and it was not taken lightly.

The United States, seeing its very survival at stake, was suddenly shocked out of complacency. We needed to get satellites aloft. We needed to improve our weapon delivery systems. We needed new technology on a lot of levels. We rose to the challenge by greatly increasing government spending on scientific research and development, and creating science-based agencies including the new National Aeronautics and Space Administration (NASA).

The quality of our children's education was addressed. Public schools received an influx of cash with new emphasis on math, science, and technology. The National Defense Education Act encouraged students to go to college and study math and science, by paying their tuition and expenses.

We recognized a threat and rose to it, applying resources on all levels as needed. After all, our integrity as a nation—perhaps our survival—was at risk.

Flush with the necessary funds, and the support of the government and the American people, NASA did manage to meet Kennedy's goal and land men on the moon in 1969.

Our first moon walkers placed a plaque on the surface stating we had come "in peace, for all mankind." But they also unfurled the American flag lest anyone be confused as to who had accomplished this feat, and was the more technically advanced country.

Human history is one of coming together and devoting staggering amounts of time and resources to projects considered essential to our beliefs and our lives. Under threat we have responded decisively and dramatically. The Great Wall of China, intended to keep out Manchurians; the dikes of Holland, to keep out the ocean; and the massive dedication of resources to develop the atom bomb to keep out the Nazis all presented tremendous challenges but were seen as necessary in light of the peril we faced.

We are at such a time again, right now. Our survival is threatened. Today it is not some external threat from "others." We don't have the images of vast goose-stepping armies marching down the Unter den Linden to shake us out of our complacency. This time it is our own behavior, which, admittedly, is harder to come to terms with.

Our parents did not know that they were damaging the earth's ecosystem with their reliance on fossil fuels. They did not know they were making life more tenuous for us. But we know it. We also know we don't need to; there is plenty of good, clean energy at hand. There are already proven technologies and more on the horizon that can make this energy available, and continue our high standard of living for even more people. There is every reason to believe that this can propel us into a better world for everyone.

But today we need the courage to face the dangers of our reality, the will to move out of our repose, and the creativity to move forward into uncharted territory with the same sense of determination as those ancient builders of the Hagia Sophia or the pioneering space travelers of the sixties. Just as we look back on them and appreciate their accomplishments, so will our grandchildren be looking back on us, hopefully with the same admiration.

Frequently Asked Questions

Q: *CO_2 and the other greenhouse gases make up only a small part of our atmosphere. How can they possibly have that great an effect?*
A: Be grateful for small favors. GHGs are indeed trace, being measured in parts per million and parts per billion. That, however, is enough; 180 to 300 ppm has been adequate to keep our earth at a comfortable living temperature, rather than the icy world that would be expected otherwise. The higher levels correlate with higher temperatures both historically and currently.

Our sister planet, Venus, which has an atmosphere that is primarily carbon dioxide, holds huge amounts of heat in the atmosphere and, not that anyone has tried it, we are told lead would melt on its surface.

Q: *Since plants need CO_2 for photosynthesis, isn't more CO_2 in the atmosphere a good thing—more plants? And wouldn't that help because bigger plants will absorb more carbon dioxide?*
A: This is a good question and actually fairly complicated.

- All plants alive today have adapted to and evolved in an atmosphere with between 180 and 300 ppm CO_2. This suggestion assumes that if a certain amount of something is good, a lot more must be better. As a rule, this is not true.
- That said, there has been some research demonstrating that some plants grown in a very high CO_2 environment will grow larger. Not all. Slightly more vigorous growth has been seen in poison ivy, of all things. However, other factors will likely limit or even reduce growth, including drought and changing climate without time for plants to adapt. Additional heat helps some plants, hurts others. And ultimately plants seem to adapt to the larger amounts of carbon dioxide, and, following their genetics, return to

their previous structure. The following website provides a good overview of the pluses and minuses for plants in higher CO_2 environments: https://www.newscientist.com/article/dn11655-climate-myths-higher-co2-levels-will-boost-plant-growth-and-food-production/.

- The problem with the quantity of photosynthesizers on the planet is not lack of carbon dioxide but lack of plant life due to deforestation as well as the pollution of the oceans that is reducing sources of sea photosynthesis.
- Any possible gains in photosynthesis would be insignificant compared to the negative consequences of global warming.

Q: *Water is a greenhouse gas—why aren't we trying to reduce the water in the atmosphere?*
A: The amount of water (humidity) is highly variable. Water remains in the atmosphere for a relatively short time, unlike excess carbon dioxide, which can remain for centuries. Global warming has increased the overall water in the atmosphere, and will continue to do so because increasing temperatures cause increasing evaporation. The only way to reduce or stabilize this GHG, water, is to reduce or stabilize temperature. That takes us back to CO_2 emissions.

Q: *Recent data shows a cooling trend; why isn't anyone talking about that?*
A: There has not been a cooling "trend," just natural fluctuations. Earth's temperatures are variable: some areas and some years are warmer or cooler than others; this has always been the case. However, the overall trends are clearly up. Each of the last three years broke the previous record for temperature, 2016 being the hottest year on record since records have been kept.

The year 2017, for various natural reasons, is projected to be slightly cooler, but this does not mean that warming is slowing. These are normal fluctuations.

Q: *Burning of fossil fuels is adding 29 gigatons of CO_2 to the atmosphere every year. But that is tiny compared to the 750 gigatons that nature puts into the atmosphere. Isn't nature the real culprit?*
A: Remember, it is a cycle. Nature also removes roughly 750 gigatons of CO_2 yearly, primarily through photosynthesis. The 29 gigatons (actually, some put current numbers at closer to 40 gigatons[1]) is more than what the natural carbon cycle can remove. Hence, it is collecting in our atmosphere and being absorbed into our oceans. The level of atmospheric CO_2 is increasing each year as we put more into the atmosphere than can be removed through natural processes.

Q: *Wouldn't it be better to delay action until we are sure?*

A: Sure of what? We are sure that anthropogenic emissions of carbon dioxide, and other greenhouse gases, are building up in the atmosphere and oceans, and will cause serious problems for ourselves and future generations. We are not totally sure how fast the temperature will rise, how fast the oceans will rise, nor can we predict precisely the consequences—although we have some very good ideas.

But to wait before taking action is a bit like being parked on the railroad tracks with a train bearing down on you and not wanting to move until you are certain how extensive the damage will be to your car.

Q: *Why do so many climate scientists refuse to acknowledge the other side, and insist global warming is "settled science"?*

A: Because global warming *is* settled science. There is no "other side" that is supported by evidence. Science is always open to new findings. If the discussion is about verifiable evidence and its implications, there will always be a ready audience. However, opinions based on incomplete data or, more ominously, efforts to confuse and delay, as was done with the science about smoking for forty years, is not helping the problem. We don't have forty years.

When searching for truth it isn't beneficial to continue age-old arguments about whether the world is flat, or left-handed people are inherently evil, or the Holocaust did not happen, or gravity does not exist, or the moon is made of green cheese, unless there is actual evidence that would support any of those ideas.

Marc Moreno, who has made a career confusing scientific issues, said in the documentary *Merchants of Doubt*: "I'm not a scientist, although I do play one on TV occasionally. Ok more than occasionally. . . . Gridlock is the greatest friend that a global warming skeptic has because that's all you really want. There is no legislation we are championing. We are the negative force. We are just trying to stop stuff."

If Mr. Moreno, or anyone, has new data or information to bring to the table, it is important that they be included. But if the goal is not truth but "gridlock," and the effort to "stop stuff," we just don't have the time.

Appendix A

Units of Measuring Energy and Power

Energy: The force used multiplied by the distance moved.

The joule (J) is the SI (International Standard) unit and is equal to a force of 1 newton applied over a distance of 1 meter. One newton is equivalent to the weight of about an apple. A meter is a little more than a yard.

British thermal unit (Btu) = 1055 J
Foot-pound = 1.4 J
Horsepower-hour = 2.7 MJ
Gasoline-gallon equivalent = 2.7 MJ
Kilowatt-hour = 3.6 MJ
Therm (100 ft^3 natural gas) = 1.055 MJ
(Outside the United States, natural gas is sold in gigajoules.)
calorie (lowercase c) = 4.9 J
Calorie (uppercase C) (1,000 calories, used in food) = 4.9 KJ

Power: The rate of energy usage. It is the amount of energy used in a given time.

The watt is the SI unit for power and is equal to 1 J per second. (J/s)
Horsepower = 33,000 foot-pounds per minute = 746 W

Appendix B

Electricity Terms

- **Coulomb** = 62.5 billion, billion electrons. (6.25×10^{18})

  We use electrons to transfer energy. But electrons are ridiculously small. Rather than talk about the energy carried by a single electron (which is also ridiculously small) we use a coulomb. A coulomb is 62.5 billion, billion. Like a dozen, it is only a number, a shortcut.
- **Ampere (Amp, Amperage)** = 1 coulomb per second.

  This is a measure of how fast electrons are traveling through a wire, the "current." One amp means 62.5 billion, billion electrons are passing by every second.
- **Volt (Voltage)** = 1 joule per coulomb.

  A volt is a measure of the amount of energy the electrons are carrying. A volt means a coulomb of electrons are carrying a single joule of energy.
- **Resistance** refers to how much a material "resists" the flow of electrons. This slows the flow of electrons. It is at "resistors" that electrons drop off their energy load.

  Your toaster has material that resists the flow of electrons. As electricity flows through these thin materials, energy is given up and they glow and give off heat. This is what toasts your toast.

  Sometimes the resistor can be a coil wire that becomes an electromagnet when electricity flows through it. If it is in a magnetic field, the coil will turn—against the resistance of the field. This is how we get electricity to turn things.

Notes

PREFACE

1. Hobbes, Thomas. *Leviathan or The Matter, Forme & Power of a Commonwealth, Ecclesiastical and Civil*. London: Cambridge University Press, 1904.

13. SO WHAT?

1. James Hansen et al., "Ice Melt Sea Rise and Superstorms: Evidence from Paleoclimate Data, Climate Modeling, and Modern Observations that 2°C Global Warming Could Be Dangerous," *Atmospheric Chemistry and Physics*, March 2016, http://www.atmos-chem-phys.net/16/3761/2016/acp-16-3761-2016.pdf.
2. John Weier, "Mapping the Decline of Coral Reefs," NASA Earth Observatory, March 12, 2001, https://earthobservatory.nasa.gov/Features/Coral/.
3. Anthony Costello et al., "Managing the Health Effects of Climate Change." *Lancet* 373, no. 9676 (May 2009), http://thelancet.com/climate-and-health.
4. "DoD Releases Report on Security Implications of Climate Change," *DoD News*, Defense Media Activity, July 29, 2015, https://www.defense.gov/News/Article/Article/612710.
5. Dr. James Hansen, "Remarks in the National Assembly of France" (Discussion, French National Assembly), Paris, France. May 12, 2010.

15. THE TROUBLE WITH SCIENTISTS

1. *Merchants of Doubt*, directed by Robert Kenner. Sony Pictures Classics, 2015, DVD.
2. Keith Wagstaff, interview with Neil deGrasse Tyson, "The Future of U.S. Space Exploration After Curiosity." Time.com. August 1, 2012. http://techland.time.com/2012/08/01/neil-degrasse-tyson-on-the-future-of-u-s-space-exploration-after-curiosity.

16. FROM THE DAWN OF HUMANITY UNTIL THE INDUSTRIAL REVOLUTION

1. Joshua Hammer. "Finally, the Beauty of France's Chauvet Cave Makes its Grand Public Debut," Smithsonian.com. April, 2015. http://www.smithsonianmag.com/history/france-chauvet-cave-makes-grand-debut-180954582.

2. Jim Knight. " A Brief Look at History of Coal." Scribd. Texas Clean Energy Project. https://www.scribd.com/document/37400348/A-Brief-Look-at-History-of-Coal.

20. TURBINE-FREE ELECTRICITY

1. "Major Land Use," United States Department of Agriculture Economic Research Service, December 19, 2011, https://www.ers.usda.gov/data-products/major-land-uses/.

23. THE THIRD RAIL SOLUTION

1. "New Study Finds that 40% of Pregnancies Worldwide Are Unintended." *Guttmacher Institute*. September 17, 2014. https://www.guttmacher.org/news-release/2014/new-study-finds-40-pregnancies-worldwide-are-unintended.

24. WHY ARE WE SLOW TO ACT?

1. Michael Lloyd "Video: Harry R. Truman, folk hero of Mount St. Helens," *The Oregonian/OregonLive*, May 14, 2010, http://www.oregonlive.com/environment/index.ssf/2010/05/video_harry_r_truman_folk_hero.html.

2. Margaret Thatcher, "Speech to United Nations General Assembly (Global Environment)," November 8, 1989, *Margaret Thatcher Foundation*, http://www.margaretthatcher.org/document/107817.

3. George Bush, "Remarks to the Intergovernmental Panel on Climate Change," February 5, 1990, *The American Presidency Project*, http://www.presidency.ucsb.edu/ws/index.php?pid=18117&st=climate+change&st1.

4. Nancy Pelosi and Newt Gingrich, "Ad on Climate Change," *We Can Do This*. https://www.youtube.com/watch?v=6tW590-BH5c.

5. Neil DeGrasse Tyson, Tweet, August 13, 2014, https://twitter.com/attn/status/499651411316310016?lang=en.

6. Joe Ryan, "Big Oil Unexpectedly Backing Newest Non-Fossil Fuels," *Bloomberg Markets*, May 2016, https://www.bloomberg.com/news/articles/2016-05-10/big-oil-unexpectedly-backs-newest-non-fossil-fuels.

25. DISRUPTIONS AND JOBS

1. "DOE Releases Second Annual National Energy Employment Analysis," Energy.gov, January 2017, https://energy.gov/articles/doe-releases-second-annual-national-energy-employment-analysis-0.
2. Robert Bacon and Masami Kojima, "Issues in Estimating the Employment Generated by Energy Sector Activities," World Bank, June 2011, http://siteresources.worldbank.org/INTOGMC/Resources/Measuring_the_employment_impact_of_energy_sector1.pdf.
3. Joshua M. Pearce, "What if All U.S. Coal Workers Were Retrained to Work in Solar?" *Harvard Business Review*, August 08, 2016, https://hbr.org/2016/08/what-if-all-u-s-coal-workers-were-retrained-to-work-in-solar.

26. "FREE MARKET"

1. Matthew L. Wald, "Fossil Fuels' Hidden Cost is Billions Study Says," *New York Times*, October 19, 2009. http://www.nytimes.com/2009/10/20/science/earth/20fossil.html.
2. Milton Copulos, "Averting Disaster of Our Own Design," testimony before the Senate Foreign Relations Committee, *EV World*, March 2006, http://www.evworld.com/article.cfm?storyid=1003.
3. "IMF Survey, Counting the Cost of Energy Subsidies," International Monetary Fund, July 2015, https://www.imf.org/external/pubs/ft/survey/so/2015/NEW070215A.htm.

27. LOOKING FOR THE POSSIBLE

1. John Schwartz, "A Conservative Climate Solution: Republican Group Calls for Carbon Tax," *New York Times*, February 7, 2017, https://www.nytimes.com/2017/02/07/science/a-conservative-climate-solution-republican-group-calls-for-carbon-tax.html?_r=0.
2. Chris Mooney and Juliet Ellperin, "Senior Republican statesmen propose replacing Obama's climate policies with a carbon tax," The *Washington Post*, February 8, 2017, https://www.washingtonpost.com/news/energy-environment/wp/2017/02/07/senior-republican-leaders-propose-replacing-obamas-climate-plans-with-a-carbon-tax/?utm_term=.e8dd34a5a518.
3. James Hansen, "Why I must speak out about climate change" TED Talk, March 2012, https://www.ted.com/talks/james_hansen_why_i_must_speak_out_about_climate_change.

28. CAPITALISM

1. Sam Geil, Interview with the one of the authors. April, 2017.
2. Sam, Geil "Business Sustainability, Harmonizing Business for Long Term Success." *Geil Consulting*, Power Point Presentation, 2017.
3. Rana Foroohar, "Saving Capitalism, American Capitalism's Greatest Crisis," *Time*, May 11, 2016, http://time.com/4327419/american-capitalisms-great-crisis/.
4. Joshua S. Hill, "World's Largest Oil Company, Saudi Aramco, Apparently Mulling $5 Billion Renewable Investment," *Clean Technica*, January 31, 2017, https://cleantechnica.com/2017/01/31/worlds-largest-oil-company-saudi-aramco-apparently-mulling-5-billion-renewable-investment/.

29. WHAT LIES AHEAD?

1. Louise, Burfitt-Dons, Speech on Global Warming, Institute of Physics, London, June 2008.

30. ENDING THE GAIAN WAR

1. *Merchants of Doubt*, Directed by Robert Kenner, Sony Pictures Classics, 2015, DVD.

FREQUENTLY ASKED QUESTIONS

1. Phil Plait. "Did I Say 30 Billion Tons of CO_2 a Year? I Meant 40." *Slate*. August 20, 2014. http://www.slate.com/blogs/bad_astronomy/2014/08/20/atmospheric_co2_humans_put_40_billion_tons_into_the_air_annually.html.

Glossary

Atom	The smallest unit of a chemical element. There are 92 kinds of naturally occurring atoms, representing 92 different natural elements.
Biodiversity	Variety and variance of living things in an ecosystem.
Biomass	Organic matter (usually plant matter) used in fuel.
Chemical potential energy	Energy stored within molecules, in the bonds between atoms.
Chlorofluorocarbons	Man-made chemicals used in aerosol sprays and refrigeration. They are greenhouse gases and also cause chemical reactions that destroy the protective ozone layer in the stratosphere.
Combustion	A chemical reaction where organic matter is "burned"—it is combined with oxygen and gives off energy, carbon dioxide, and water.
Conservation of energy	The scientific principle that states that energy cannot be created or destroyed, it is just converted to other forms.
Desalinization	A distillation process that creates freshwater from salt water by removing the salt.
Distillation	A process where materials are removed from water by boiling, evaporation, and cooling—leaving impurities behind.

Ecosystem	A grouping of living things, plants, animals, microorganisms, and their environment that are interrelated and dependent on each other. It can be very small, or the whole world.
Electromagnetic radiation	Energy-carrying waves that travel through space at a speed faster than anything else can travel. There is a spectrum of these waves from very low-frequency, low-energy photons like radio waves, up to high-frequency, high-energy photons called gamma waves. Visible light is a small part of the spectrum. Infrared are below the visible spectrum; ultraviolet are just above.
Elements	The most basic substances, out of which everything else is made. There are 92 naturally occurring elements, each having one more proton in the nucleus than the one before it. They range from hydrogen, a very light gas with 1 proton, to uranium, a very heavy solid with 92 protons.
Frequency	How many times something happens in a second. Wave frequency would mean how many times a wave goes through its full motion in a second.
Giga	Billion.
Gravitational potential energy	Potential energy that an object has by virtue of its place in a gravitational field. The amount of energy it has is a combination of the weight of the object and its height.
Greenhouse gases	Gases that hold heat in the atmosphere. They are molecules of 3 or more atoms. The most abundant ones are water, carbon dioxide, methane, nitrous oxide, ozone, and chlorofluorocarbons. All occur in nature, except chlorofluorocarbons, which are man made, and all are increasing in the atmosphere.
Hydroelectricity	Electricity produced using the kinetic energy of water as it flows to the sea in rivers.
Kilo	One thousand.
Kinetic energy	The energy of motion. Anything that is moving has kinetic energy.
Luminiferous ether	A material that many scientists once believed filled space. It was considered to be rigid and massless and

	light traveled through it. It doesn't exist. Light doesn't need a medium to "wave" because it is also a stream of particles called photons.
Mega	One million.
Methane	A greenhouse gas—the primary component of "natural gas." It is used as fuel, and releases carbon dioxide and water, but less carbon dioxide for an amount of energy than coal. However, leaks of methane in the retrieval and burning make its preference to coal questionable.
Molecule	A combination of atoms that exists as a stable unit.
Natural gas	A gas used as fuel, composed primarily of methane.
Nitrous oxide	A greenhouse gas.
Nuclear fission	The breaking apart of large nuclei, with a chain reaction, that produces large amounts of energy. This process is done in nuclear power plants and nuclear bombs.
Nuclear fusion	A process of forcing two hydrogen nuclei together to form helium, releasing huge amounts of energy. This is the energy running the stars and in hydrogen bombs.
Nuclear potential energy	Energy stored in the nucleus of the atoms.
Peat	A fossil fuel, a precursor to coal. It is often burned for fuel.
Photosynthesis	A process by which green plants, and other organisms, convert CO_2 and water into sugar, giving off oxygen as a waste product.
Photovoltaic cells	Panels that can convert light into current electricity.
Phytoplankton	Microscopic aquatic plants that produce large amounts of oxygen for the world and are the basis of the food chain in the ocean.
Potential energy	Stored energy. Chemical, nuclear, gravitational are three important ones relating to global warming.
Radioactive decay	Natural process by which unstable nuclei, usually in the heaviest elements, give off particles to become more stable. There are three basic types: alpha, which ejects two protons and two neutrons, which

	become a helium nucleus; beta, which ejects an electron, which converts a neutron to a proton; and gamma, which ejects a powerful photon. Gamma is the most dangerous of these since neither alpha nor beta particles can penetrate very deeply. On the other hand, if ingested, any of these can cause damage to cells and organs.
Respiration	The process that is the opposite of photosynthesis. Sugar (and other organic molecules) are broken apart into carbon dioxide and water, releasing the energy captured in photosynthesis.
Stratosphere	The atmosphere is layered. The bottom layer in which we live is called the troposphere. The ozone layer, which protects us from ultraviolet radiation, is in the stratosphere, which is the next layer up.
Tera	One trillion.
Transmute	When a nucleus gains or loses one or more protons and becomes a different element.
Troposphere	The lowest layer of the atmosphere.

Annotated Bibliography

BOOKS

There is a plethora of books and media on climate change. Some are lengthy and highly technical. Here are a few that go into more depth on various aspects of the problem but are generally reader friendly.

General In-Depth Information about Climate Change and Global Warming

Klein, Naomi. *This Changes Everything: Capitalism vs. the Climate*. New York: Simon & Schuster, 2014.

A comprehensive and well-researched book. Ms. Klein challenges us to look at our economics and the need to change basic attitudes in an era of climate change and environmental degradation.

Mann, Michael E., and Lee R. Kump. *Dire Predictions: Understanding Global Warming; The Illustrated Guide to the Findings of the IPCC*. 2nd American ed. New York: DK, 2015.

This book by Mann and Kump is an overview of the 5th assessment of the IPCC, written for the general public. Good pictures and graphs. Recommended for a comprehensive but readable assessment of the current climate science.

Weart, Spencer. *The Discovery of Global Warming*. Cambridge, Mass.: Harvard University Press, 2008.

Part history, part science, part detective story, this book makes science and scientists real and understandable. Highly recommended.

A Brief Overview

Goodwin, Dean, and Joe Lee. *Climate Change for Beginners.* 3rd ed. Danbury, Conn.: For Beginners, 2012.

Short, easy to read, and covers the basics. Clever drawings.

Books Addressing Confusion and Specific Questions about Climate Change

Bennett, Jeffrey. *A Global Warming Primer: Answering Your Questions about the Science, the Consequences, and the Solutions.* Boulder, Colo.: Big Kid Science, 2016.

Dr. Bennett writes in question-and-answer format, addressing many of the concerns and confusions about climate change. Easy to read but authoritative and accurate. If you only read one book, this might be your choice.

Darling, Seth B., and Douglas L. Sisterson. *How to Change Minds about Our Changing Climate: Let Science Do the Talking the Next Time Someone Tries to Tell You . . .* New York: Experiment LLC, 2014.

Solid science to counter the confusion that is prevalent in society today. Easy, friendly read, with some humor.

Exploring Why We May Be So Slow to Take Global Warming Seriously

Marshall, George. *Don't Even Think about It: Why Our Brains Are Wired to Ignore Climate Change.* New York: Bloomsbury, 2014.

An important look into a lot of the psychological factors that cause us to look the other way when faced with serious problems, and how we can overcome our reluctance and illusions, to take effective action.

Oreskes, Naomi, and Erik M. Conway. *Merchants of Doubt: How a Handful of Scientists Obscured the Truth on Issues from Tobacco Smoke to Global Warming.* New York: Bloomsbury, 2010.

A historical and scientific look at the effective campaigns that have been run to confuse the science, slow action, and mislead the public on issues including the health effects of smoking, coal smoke, acid rain, the damage to the ozone layer from CFCs in the stratosphere, and now, global warming. An important read relevant to the state of democracy in America today.

Seethaler, Sherry. *Lies, Damned Lies, and Science: How to Sort Through the Noise around Global Warming, the Latest Health Claims, and Other Scientific Controversies.* Upper Saddle River, N.J.: Pearson Education, 2009.

This book addresses the larger problem of misinformation about scientific information.

Emphasis on Looking toward the Future

Bloomberg, Michael, and Carl Pope. *Climate of Hope*. New York: St Martin's, 2017.

Two very different but highly successful individuals show how businesses, cities, and individuals can be a bold force in solving the climate crisis. They demonstrate the knowledge, experience, and passion needed to come to solutions. An important read.

Flannery, Tim. *Atmosphere of Hope: Searching for Solutions to the Climate Crisis*. New York: Text Publishing, 2015.

An Aussie's perspective, this book describes the science of climate change and addresses many of the innovations both current and in development. The author offers a real sense of hope for our future.

Nye, Bill. *Unstoppable: Harnessing Science to Change the World*. Edited by Corey S. Powell. New York: St. Martin's, 2015.

Writing in his usual conversational and entertaining style, Nye emphasizes solutions for the future from conservation to high tech. He explains science without resorting to "science-talk."

AUDIOVISUAL RESOURCES

There are a number of documentaries addressing global warming, the science, and the solutions. Here are a few. Most are available online, from Netflix and others.

Chasing Ice. New Video Group, National Geographic, 2013.

This film follows a film crew as they document the melting of glaciers in the Northern Hemisphere. 75 minutes.

Climate Change, Lines of Evidence. National Academy of Sciences, 2012.

Good summary of the evidence for global warming. 25 minutes.

An Inconvenient Sequel: Truth to Power. Participant Media, July 28, 2017.

A sequel to Al Gore's *Inconvenient Truth* from 2007. An updated visual presentation of the realities of climate change and the many people who are working on solutions. 100 minutes.

The Island President. First Run Features, 2012.

A documentary, a primer on climate change, and a political thriller. 101 minutes.

Merchants of Doubt. Sony Pictures Classics, 2015.

Based on the book of the same name, entertaining, informative, and sobering. 96 minutes.

"The World Set Free." Episode 12 of the updated *Cosmos* series. Cosmos Studios, Fuzzy Door Productions, 2014.

An excellent review of the topic with great graphics. Presented by Neil deGrasse Tyson. 45 minutes.

Years of Living Dangerously. Showtime, 2014.

A nine-part series on climate change featuring a host of celebrities, scientists, and citizens. This will take you nine hours.

About the Authors

Jeffrey B. Loehr is a strategy advisory and growth specialist, helping businesses understand and manage uncertainty in the future. He has strategic management experience in industry and as a consultant across a range of industries including automotive, technology, natural resources/mining, and consumer products. He writes regularly online and for publications.

Climate change is becoming a major issue and opportunity for business, and Jeff is passionate about understanding, managing, and improving the situation while embracing the implications.

He has a bachelor's degree in economics from the University of California and an MBA from the Anderson School of Business at UCLA. He is also a yoga instructor in White Plains, New York, where he lives with his wife, Magdalena, and two children.

Margaret K. Loehr has a degree from the University of California at Davis in nutrition and a master's in public health from UCLA. She has taught science and nutrition in clinical settings, at the community college level, and, most recently, as a high school teacher explaining the scientific concepts of biology, chemistry, and AP physics.

Her dedication to education about climate change comes not only from her concerns for her children and grandchildren but from a sense of responsibility she has for the many young people she has taught and the world they are inheriting.

She and her husband, Burt, are docents for the Crocker Art Museum in Sacramento, California.

CPSIA information can be obtained
at www.ICGtesting.com
Printed in the USA
BVOW09*0346061017
496362BV00002B/4/P

9 781475 835816